Dedicated to
Owen Wright and
the late Peter Wright

About the Authors

Rob Davidson *was born in 1920 in Cape Town, South Africa. He graduated from the University of Witwatersrand with an honours degree in bioecology at a time when plant and animal ecologists were trying to unify the two disciplines. After working at the Veterinary Research Laboratory in Kenya, he returned to the university to lecture in botany and supervise postgraduate research in ecology and pasture management.*

In Australia since 1961, he worked in the CSIRO, researching agricultural ecosystems. Rob has published numerous scientific papers and written newspaper and magazine articles encouraging farmers to be more aware of the environment.

Steve Davidson *was born in 1954 in Johannesburg and came to Australia with his family at the age of seven. He graduated from the University of New England with a masters degree in ecology and, after working with the NSW Parks and Wildlife Service, he joined the CSIRO in 1982 as a science writer. Steve has produced some 150 articles on agriculture, biology and environmental science and maintains a keen interest in wildlife photography.*

The contributors to chapter 14 all live and work in the States in which the success stories took place. They may be contacted through AGPS Press, GPO Box 84, Canberra ACT 2601.

BUSHLAND ON FARMS
DO YOU HAVE A CHOICE?

ROB AND STEVE DAVIDSON

with contributions from
Elizabeth Bond, Judy Frankenberg, Rhonda Lyons, Prue Powell,
Jill Reading, Julianne Venning and Peter Voller

An AGPS Press publication
Australian Government Publishing Service
Canberra

ISBN 0 644 14444 0

National Library of Australia Cataloguing-in-Publication data:

Davidson, Rob, 1920–
Bushland on farms: do you have a choice?

Bibliography
ISBN 0 644 14444 0.

1. Agricultural ecology – Australia. 2. Agriculture – Environmental aspects – Australia. 3. Agricultural pests – Biological control – Australia. 4. Agricultural pests – Integrated control – Australia. 5. Agricultural pests – Cultural control – Australia. I. Davidson, Steve, 1954– . II. Title.

630.994

Designed by Ross Pulbrook; figures by Fiona Tout & Marcus Flux, AGPS Design Studio
Edited by Sharon Nevile, AGPS Press
Typeset by Signature Jer 31: 33 Canberra
Printed for AGPS Press by Pirie Printers Pty Ltd, Fyshwick, ACT 2609

Foreword

By the Ambassador for the Environment, Sir Ninian Stephen

In a world which suffers, still, the ravages of poverty and famine, we do indeed live in a bountiful land. For more than 200 years our farmers and graziers have ensured that we Australians would never experience hunger. At the same time, they continue to produce the surplus which provides us with a major source of export income.

Pictures of rural Australia tend to be etched on the national psyche, not as a symbol of Arcadian idyll but as a symbol of the economic prosperity and the wellspring of those things of culture which we identify as peculiarly Australian. Yet a peep behind the picture will soon tell us that all is not well beyond the farm gate. The national prosperity that our rural industries have given us has come at a price, sometimes a very high price. Every year we lose more of our productive land to erosion, salinity, acidification, decline of soil structure and to pests such as rabbits. We know now, of course, what aboriginal Australians have known for centuries — that our land can be as harsh and unforgiving of abuse as it can be bountiful with care.

This is a wonderful book. At a time when we might all despair about the future, here we read of hope — hope given of trial and error, of failure and, indeed of despair and real emotional struggle. We read of the methods and the successes, not in the words and language of experts but of ordinary rural people who can talk through these pages to so many other ordinary rural people whose commitment will be critical if we are to make a truly ecologically sustainable Australia for the future.

The message of this book is delivered quietly but clearly; we who depend upon our farms and grazing lands for sustenance and economic prosperity must recognise that they are created from whole functioning ecosystems. Nature is remarkably resilient but the strain is showing and it is we who are primarily imposing that strain. Here you will find people who have come to realise the limits to which nature can be taken, who know that one process affects another and that activities by man can have far reaching and unexpected, sometimes undesirable, consequences. They have learned that the retention and maintenance of native floral and faunal communities can be just as important to production as irrigation, fertilisation and insecticides. Their future, and the future of the majority of us who live in our great sprawling cities, is ultimately linked to healthy natural systems and an appreciation of the agreement we must strike with nature to maintain that health.

And for that which is saved for nature by landholders like these, nature will repay handsomely in dividends of pest control, watershed integrity and water quality maintenance, erosion and salination prevention and preservation of the biological diversity which has always been the source of much that nourishes us. We can thank them, too for helping stem the fragmentation and drainage of natural habitat which continues to put many species at risk.

I am extremely pleased to be associated with Bushland On Farms. *In language that we can all understand, it presents to us the 'whole' complex picture of the way our land is and the way it could be in the future. It offers us hope as we strive to better understand it, avoid the mistakes of the past and find the accord with nature that will enable future generations to inherit a world worth inheriting.*

Contents

Illustrations

Figures

TABLES

PHOTOGRAPHS

Colour Plates

Acknowledgments

A great number of people have helped us tell a good story — their story in parts. Owen Wright at 'Wongwibinda', the late Peter Wright at 'Lana' and Jillian Oppenheimer at 'Ohio' have shared their memories of pioneer families. The encouragement given by these families, about whose farming activities we are writing, makes it possible to be frank in suggesting a change of attitude. John and Cicely Fenton, Bill McClymont, Susan Campbell, Tim Cox, Vince Wolfe and Mary Chapman have given encouragement and ideas over many years.

Special thanks go to Milton Moore, Doug Waterhouse and Max Whitten, and to the late Phil Carne for reading and commenting on the manuscript. Other colleagues have helped in providing data and commenting on their specialist topics, particularly the late Harry Frith, Bunny Fennessy, John Leigh, Hal Heatwole, Tony O'Farrell, Hugh Ford, Robert Boyd, John Williams, Gus Campbell, David Curtis, Reg Roberts, James Ridsdill Smith, Graeme Baker, Peter Haskins, Steve Goldsmith, Suzie Gaynor, Lindsay Prior, Andy Andrewartha and Tim Austin. Discussions with them have been most helpful.

Many others have read the manuscript and suggested changes in style and content. Felicity Leigh, Joan and Kate Boyd, Judith Wood, Val and John Wiseman, Max Woods, Brian Guilfoyle, Bill Willoughby, Barry Hall and Geoff Quinn are thanked for their critical editorial comment. Special thanks to the late John Turner who encouraged us by meticulously going through the manuscript and making many useful suggestions. We thank Marg Henderson and Cheryl Chant for preparing the manuscript. The staff at AGPS Press have also made an invaluable contribution.

The Department of the Arts, Sport, the Environment and Territories has supported publication in the public interest. In particular, the Environmental Planning Branch of the Environment and Conservation Policy Division of the department facilitated publication and Ian Gorrie, then an officer of the branch, is thanked for encouragement and administrative assistance. Departmental funds used came from the Environment Education and Information Unit.

We acknowledge the kind assistance of the Land and Water Resources Research and Development Corporation in offering financial support to this publication.

The map appearing as figure 19 is Crown Copyright and has been reproduced with the permission of the General Manager, Australian Surveying and Land Information Group, Department of Administrative Services, Canberra ACT.

The success stories forming the majority of chapter 14, and the photographs associated with each story, are reproduced with the permission of their authors. Unless otherwise acknowledged, all other photographs appearing in this book are the property of its authors.

RLD and SJD

February 1992

INTRODUCTION

Across the world uneasiness about the excessive use of insecticides in homes, in gardens and in agriculture is growing. In Australia, the Senate Standing Committee on Agriculture and Veterinary Chemicals (1991) recommends that research and development and responsible farming practices should aim to reduce the use of pesticides wherever possible.

Yet our growing population needs to be fed, sheltered and clothed. Also the livelihood of farmers depends on economically justifiable control of crop pests. Where do you stand in this dilemma? People who have to make the decisions — the graziers, farmers and gardeners — have very little information about natural controls. After all, who is going to promote natural controls when there is nothing to sell?

The extension services of Departments of Agriculture, the pest control advisers, agronomists, horticulturists or nurserymen could spell out to growers the details of integrated pest management: how to use nature's controls to the full, with the minimum of insecticides. But you only ask for advice when your crop is being attacked by some pest. You would not be impressed if you were advised to grow trees to encourage magpies, and plant flowering shrubs for wasp parasites of insect pests. Yet the loss of habitat for birds, wasps, robber flies, spiders, lizards and all the other biological controls of pests is as important to the decline of nature's way of pest control as the toxic effects of chemicals on these creatures.

During our scientific studies it became clear to us that we were taking too narrow a view. While we were concerned with scientific details, the habitat of natural controls was being destroyed by the same agricultural development we tried to serve. In this book, control of pasture pests has been used as an example of the ecological approach to pest control. The essential habitat requirements of natural controls of pasture pests make fascinating reading. Where do the wasps find nectar? Where do the birds find insects to nourish themselves year-round and feed their ravenous chicks? What influence does climate have on all these insect populations? Research carried out at great expense provided most of the answers to these questions, and the results persuaded New England graziers that insecticides were not necessary. A firm, commonsense decision prevented the potential problem of pesticide residues in animal products and avoided an enormous waste of money.

Weather, disease and naturally occurring predators and parasites of insect pests collectively exert an awesome control over insect pest populations, and in most seasons these natural controls are so effective that farmers and gardeners need not spray with insecticides as either a preventive or a curative measure. Toxic sprays need only be used as a last resort when natural methods are not working. Native bushland has a unique value on farms as a habitat for predators and parasites of insect pests. This book explains the functioning of bushland ecosystems composed of trees, shrubs, herbs, grasses and a wide variety of animals which feed on plants and on one another. Some animals find shelter in dead trees and in the litter of bark, twigs and dry leaves on the ground. Unless farmers can be persuaded to fence out livestock from relatively small areas of bushland (such as steep and rocky slopes and river banks), and manage these with understanding, their pest problems will slowly worsen.

Sadly, though, on farms around Australia natural controls of insect pests are under threat because of rural tree decline and loss of bushland due to clearing. The countryside is changing so fast that in half a lifetime it will look different even to the casual observer. What is the landholder's role in these dynamic processes? This book tries to explain what is happening. It aims to point out to farmers and townspeople what they stand to lose if native vegetation disappears. The losses would be economic (measurable in dollars and cents) and aesthetic (reducing the quality of the environment for all Australians). Natural predators and parasites of pests not only provide a free pest control service but they, like the bush they inhabit, are nice to have around.

Governments, farmers, scientists and conservation groups are all talking about the concept of sustainable agriculture. Now surely the time has come for landholders to act, and regeneration or planting of native vegetation is one of the best forms of action — **as long as farmers realise that they need diverse, living ecosystems, not just trees.**

PHOTOGRAPH 1

Tree death without replacement is an insidious form of habitat loss with dire consequences for a host of dependent species. Photo by Judy Gall, courtesy *Habitat Australia*.

We and many others have managed to achieve this simply by fencing off areas from stock to allow retention and natural regeneration of the bush, and by planting tree and shrub seedlings raised from seed gathered locally. It is not difficult — fencing is the hardest part — and it is extremely rewarding.

Much of this book is set in the tablelands of New South Wales, but the ecological principles apply generally to all farmland throughout Australia. Similar plants and animals will be found wherever bushland habitat is provided. Success stories describing regeneration and planting of bushland are presented by seven contributors from all States of Australia.

If the book dispels two common misconceptions it will serve its purpose. The first is that bushland is incompatible with modern agriculture, and the second is that an insect pest problem will worsen from year to year if the pest is not eradicated each time it is seen.

1

A Place to Roost

Fifteen years ago we had a dam built across a trickling stream graced with the name Bullock Creek, near Armidale. The story is worth recalling. Not only was there drama and humour in tampering with the powerful force of rushing stormwaters; there were also unexpected pleasures and lessons to be learned. In particular, we were reminded of how much birds and other wildlife value bushes and trees, even dead ones standing in water.

While doing a survey and preparing a plan for the earth wall, the engineer from the Water Resources Commission remarked that it was an ideal dam site because gravel and silt over the years had created a long level floodplain. A low wall would push the impounded water well upstream and the shallow water around the edges would encourage water birds. Two tributaries join the creek, and the survey showed that water would be pushed into these to give a pleasing diversity. It was necessary to dig a trench through the gravel creek bed down to bedrock and fill this with basaltic clay to provide a waterproof core for the wall. Using an experimental seismometer, a friend, John Stanley, had calculated that the riverbed had 60 cm of dry alluvium above the water table and 120 cm of wet gravel on bedrock at a depth of 1.8 m from the surface. Continuous pumping, with a backhoe digging the trench while a bulldozer filled the trench from one end and compacted the clay, was the answer to this problem. Work continued well after dark. We knew the contractor would pull out if this stage was not successful.

By the morning it was clear that the core was waterproof, and work on the wall commenced. Then a difference of opinion arose between the contractor and ourselves. We wanted to leave the large stringybark trees standing on the creek banks, but he said that trees were always removed from dam sites. 'Not this one', we replied, and our persistence won the day, except when a tree got in the way of the bulldozer. Many of the trees had been ringbarked many years ago. Furthermore, the living trees in the creekbed and perhaps the handsome stringybarks on the banks would eventually be killed by the water. The difference of opinion was really about the value of dead trees over water.

To most people they are unsightly but to many bird species they are havens of safety. But we are getting ahead of our story.

After two days the construction of the wall was complete. The bulldozer operator worked on after dark on Friday afternoon, deepening the creekbed as much as he dared. This is a very tricky job, because there is water beneath the gravel and silt and fine judgement is needed to move off just before the crust gets so thin that the heavy machine sinks into the deep mud beneath. He did a fine job until, in the darkness, it became too risky.

The young fellow wanted the weekend off in town — a rare pleasure for those whose work is mostly in the country — so he planned to return to cut a spillway on Monday. We promised to pump water out of a deep excavation upstream so that some clearing of bushes and silt could be done, at the same time raising an island for birds to occupy. But here nature took a hand. On Friday night it rained, and we went next morning to inspect the previous day's work. It started as a leisurely drive out of town in a light drizzle, but it ended in panic as we realised that a localised thunderstorm upstream had turned our trickling stream into a torrent of rushing water. Both little tributaries were running full, and we had an instant dam. But we had no spillway yet and the water level had risen to within half a metre of the top of the raw earth wall across the creekbed. The water level was inching up, and if the wall was breached untold damage would be done downstream to bridges and property. The spillway was specified to be a metre below the top level of the wall, and when built it would be big enough to divert the stormwater from the heaviest rain around the wall and back to the creekbed lower downstream. But we had put off the spillway construction until Monday.

We shall never forget the panic. We dashed back to the nearest phone booth and telephoned the contractor, who mercifully remembered where the bulldozer operator was staying in Armidale. When we reached him he was in his city clothes, looking forward to midday dinner. He had no time even to change into working clothes. He excavated the spillway on one side, and we checked the level while standing in a foot of muddy water. Then our luck ran out. While crossing the creek to construct a second spillway his massive machine sank to the axles. All efforts to extricate it with logs failed in the pouring rain (later in the week it had to be driven out with the tracks chained to railway sleepers). Our dam building was over — the dam full to the brim.

It is a dam unlike any other because we did not have time to clear the sapling eucalypts or the blackthorn bushes from the bed of the creek — shrubbery now half submerged. A burnt tree stump remains and the cormorants sit on it to dry their wings after feeding on yabbies. Wood duck scramble out on to the bare trunk of a fallen red gum. These were the first to bring their ducklings from an unknown nest, probably high in a hollow tree. Black duck also come in large numbers, and hatch out young which remain until able to fly.

Little grebes showed us the virtue of our rough and ready dam building. We had been watching these white-bellied divers over several months, charmed by the way they up-ended themselves and disappeared under water, only to reappear some distance away. Then we recognised one of them sitting on a nest, head erect, looking like a twig in a tangle of driftwood. The nest was in a blackthorn bush half submerged in the middle of the dam. We would have removed this bush had we not been robbed of the opportunity by the rising waters of that first storm. The platform of twigs and leaves was slightly raised above the water level and in the dead blackthorn it was easily overlooked. The eggs were covered with nesting material whenever the parent left the nest. Many clutches have been reared by the grebes or dabchicks in the past few years, and each time we are glad that fortuitously the blackthorn and other bushes were left standing in the water to provide safe breeding places for these shy birds. When minding young, one of the pair will make a grand display to distract attention from the youngsters in the water. With wings flapping and beating the water, the decoy will run on the water with a great flurry. At other times they seem to have become inured to us.

We were slowly coming to realise that birds will fly great distances to reach a safe place to roost at night. Little pied cormorants are seen singly on dams and streams everywhere. They have taken to roosting at night on a dead stringybark sapling standing in the deepest part of our dam. Often we have counted up to nine cormorants on this one tree, like Christmas tree decorations. They fly in, one at a time, last thing in the evening and leave again at first light.

There have been some surprises. Once a brown bittern stayed for a few days. During the day the bittern mostly makes itself inconspicuous by standing on one leg and holding its beak straight up in the air, so that it resembles the reeds in which it usually stands. Ours did this on the exposed bank of the dam, looking rather silly. We will probably not see bitterns again until reeds or rushes establish themselves on the banks because these are the special habitat of this species.

Our greatest surprise came one evening in March three years after the dam was built. At dusk a flock of 50 straw-necked ibises flew in and wheeled overhead, coming lower over the trees on the dam. They were gliding, wings stretched out and held still. Slower and slower they sailed down, close together and yet without confusion. When close to the dead trees standing in water, a few birds settled down on the thin dead twigs. The flock continued to circle, and each time a few more dropped on to the trees. We breathed a sigh as the last one settled, delighted by these unexpected visitors. Then against the darkening sky someone discerned another flock coming in quickly towards us. The perfectly coordinated approach to the trees around the dam was repeated, the powerful flight giving way to the slow graceful glide around the dam, and later towards the trees. In all, about 200 ibises settled on half a dozen dead stringybark trees, mainly on the outer branches. Then we noticed a white ibis glide down from the flock of straw-necked ibises on to a smaller stringybark sapling in the centre of the dam. It was joined by a few more white ibises, until seven perched on that small tree. All the ibises settled down for the night, with some squawking and squabbling in the dark. They flew off next morning when the sun came up. Little did we realise that this flock would increase to 1000 by June, and the evening performance would take place every day until the end of September.

Before we get lost in the superlatives which must be used to describe a nightly performance so beautiful, let us pause for a moment to consider what had happened. Ibises are encountered on the tablelands for about half of each year, feeding in pastures in flocks large and small. Between September and November they leave the tablelands and return to distant swamps to breed. Those in New England breed in the Macquarie Marshes in western New South Wales. They nest after sufficient rain provides swampy conditions, and possibly when food (such as grasshoppers and locusts) is abundant there.

They are not known to nest on the tablelands. While on the tablelands, therefore, they need only food and a place to roost at night. But why did these ibises choose a half-finished dam close to town? Was it precisely because the unexpected rain prevented the clearing of the dead trees in the water, together with our intuitive feeling that those tall stringybarks on the banks might attract waterbirds? Perhaps the security afforded by trees over water in a secluded valley was recognised by the ibises flying over. They did not roost there before the dam was built and they did not feed on the site.

Occasionally a few ibises will feed on the nearby properties. But mostly they fly 25 km or more to feed on crickets, grasshoppers and caterpillars in pastures; or insects in lucerne; or whatever they can find in newly-turned soil. They feed for an hour or two in one place, then fly off to another feeding ground. They break up into smaller flocks and disperse quite widely. But in the evening they fly to the roosting place that they know. We realised this when we were watching them coming home to roost in July. The first birds came in from the south at a quarter to five on this winter's evening. We were watching and counting them as the successive flocks came in. Easy enough to count them as they flew in long lines, or in a V-formation. But things got a bit crowded in the air above the trees. Two flocks would pass through one another, gliding in different directions, yet the control was so good that a few flaps of those great wings would separate the flocks. They wheeled around repeatedly until the last could find a perch. While we were watching the southern skyline we were surprised by a flock of several hundred flying in from high up in the northern sky. Another flock flew in very late, when it was almost dark, from the west. So no doubt these birds knew where they would roost, arriving there in separate flocks. And they sometimes left their arrival to the last possible moment. By this time the trees were so heavily laden with birds that the latecomers were roosting in tall stringybarks 30 m from the water. Towards the end of winter some birds were killed in the trees on dry land. Not many, but the sight of a pile of ibis feathers left by a predator was a reminder of the wisdom of the ibises' choice of a roosting place above water.

Throughout that season the white ibises, up to ten at a time, would slip down to the sapling in its unassailable position in deep water last thing before dark. They would first alight on the tall trees with the straw-necked ibises.

Surprisingly the white ibises also shared their small tree with a number of cormorants.

In the following year the ibises used the trees again as a roosting place, but there were no more than 300 straw-necked ibises. The first came in March and there were young birds amongst them, recognisable by the straighter beaks. Again there were a few white ibises, about seven in all, roosting on the sapling in deep water and sharing their tree with cormorants.

There are three distinct threats to ibises in modern agriculture. Firstly, the distant breeding grounds may be rendered unsuitable for nesting by reservoirs, drainage and grazing. Robert Carrick, back in 1962, discussed the conservation of breeding grounds in the Macquarie Marshes. A second threat to ibises is lack of suitable roosting places on the tablelands.The third threat is from feeding on insects that have become lethargic because of insecticides used on crops. Fortunately there is little danger to birds which feed on pasture insects, because pastures are seldom sprayed on the tablelands as the cost is prohibitive, and out of proportion to the sporadic and localised damage suffered. Arable crops such as potatoes, sorghum and sunflowers are treated with organophosphate and organochlorine insecticides and residues in soil are taken up by insects on which birds may feed.

Straw-necked and white ibises are two of the insectivorous birds that are contributing to the biological control of pasture insects. Both species of ibises feed on a very wide range of insects, spiders, yabbies and the like. Crickets, grasshoppers and locusts are the most common item in the diet of straw-necked and white ibises, although white ibises on average have more yabbies in their diet than the straw-necked ibises. From Carrick's study the highest percentages of total food items in 364 ibises were crickets (32 per cent in straw-necked, 49 per cent in white), moth caterpillars (18 per cent in straw-necked, 4 per cent in white), soil beetle larvae (6 per cent in straw-necked, 1 per cent in white), plague locusts (5 per cent in straw-necked, none in white), spur-throated locusts (4 per cent in straw-necked, 7 per cent in white) and yabbies (2 per cent in straw-necked, 11 per cent in white). These are average figures from ibises collected in the marshy breeding grounds and in dry and moist pastures. When feeding on pastures on the northern tablelands, each ibis (which weighs about 2 kg) would probably eat about 200 g of insects each day.

Remember this when you see a flock of 1000 ibises flying overhead, because when they fly to a roosting place in the evening they will have eaten nearly a quarter of a million crickets or caterpillars or curl-grubs from some lucky person's pastures during the day. (It is usual to find several hundred insects in the stomach of one ibis, depending on where it has been feeding.)

No, these ibises will not eliminate insect pests from pastures, but they are contributing substantially to reduction of insect numbers below the economic threshold of damage to pastures. Many insects feed on pastures in the tablelands, but naturally occurring diseases, predators and parasites, and extremes of weather keep their numbers down, so that these insects are not usually serious pests. We discuss some of these natural controls in the following chapters.

Ibises are magnificent birds, which delight anyone seeing them fly in a long line against the evening sky or wheeling overhead at a great height, searching for a place to land. As they rise up from the grassland where they have been feeding, the slow and powerful sweeps of their wings emphasise their huge span. Their long curved beaks are exactly right for snapping up the insects that are disturbed as the birds walk slowly through grassland, their heads hanging down. We feel privileged to have these birds on the tablelands. They are not out of harmony with the modern productive pastures throughout Australia. Rather they come because there is food for them in these improved pastures. While enjoying their presence here for six months of every year, we wonder if enough is being done to encourage these useful birds. How many flocks of ibises could the millions of hectares of pasture on the tablelands support?

The best answer we can give to these questions is that while there are insects in excess of the food requirements of the various insectivorous animals, the farming areas could support more ibises than are presently found. We have seen no census figures for ibises, but we know that in summer and autumn there is more than enough food for them. In winter the readily accessible food, such as crickets and grasshoppers, becomes scarce, but underground the scarab grubs are fat and at their largest. Cultivation of land for potatoes or just renovation of pastures with a spring-tine harrow will turn up grubs that will attract the birds in large numbers. Ibises discern these sites from the air, and they can be attracted by this means. Magpies and many other birds can also be attracted by these simple farming practices. Having attracted these birds to our properties, we should all make the effort to encourage them to stay by providing safe roosting and nesting places.

We need an understanding of the special habitat requirements of each species of bird. In a later chapter we will look at the magpies on the tablelands and see how they have fared in the new pastoral environment.

We leave the ibises with the thought that fortunately they are not deterred by the partial clearing of eucalypts. It is the simplest thing to provide secluded roosting places for ibises during their visits to the tablelands, but it is also too easy to neglect their total requirements for breeding and a year-long need for food. This food must not be poisoned. With wide-ranging species like the ibises, many people in different parts of the country are collectively responsible for their well-being.

The need for a comprehensive view is clear. Difficult decisions have to be taken, for example the retention of breeding places such as the Macquarie Marshes where there are competing interests with farmers wanting to irrigate crops. So many things that we do have unexpected consequences. When you farm, do you do so in defiance of nature or do you recognise that many natural processes such as pest control, nutrient cycling, soil improvement and erosion control can be encouraged in order to raise the profitability of farming? The field of ecology provides the only framework in biology that tries to put together what we know about plants and animals and their relationship with the environment.

2

Ecosystems and Populations

Biology has become so specialised that the projects undertaken by biologists seldom deal with more than a minute aspect of natural systems. This is an easy way for research to be done, but it is difficult to extrapolate the results and to apply them to real systems in the field. Those who try to add together the results of experimental research are frustrated by the missing pieces of the scientific jigsaw puzzle.

Ask several specialists what they see when viewing sheep at pasture and the answers will vary widely, depending upon the expertise of each one. Some will see grasses and clover, or weeds; some will see the flocking behaviour of sheep; others will watch each sheep selecting its diet. One may see crickets and grasshoppers and name them; another will count the magpies. Some will see only the dung and litter on the ground and wonder how fast it is being decomposed; others will see the furrows eroded by rivulets of water and wonder how much the soil is compacted by the trampling of sheep. Each of these specialists has something of value to add to our understanding of agriculture.

But who knows how one process affects another? The philosophy of holism teaches that the whole is greater than the sum of the parts. A simple example will illustrate the interaction of components. The growth of plants in response to nitrogen fertiliser with inadequate phosphate is quite different from the response to nitrogen when phosphate is increased (fig. l). The additional growth ('A' in the figure) is not the sum of the responses to N_2 and P_1 measured separately; it is the size of the N x P interaction.

Interactions between plants and animals and the environment are very difficult to measure in ecosystems in the field. It is a sobering thought that without a measure of these interactions or at least an estimate of them, there will be no precise and comprehensive understanding of ecosystems.

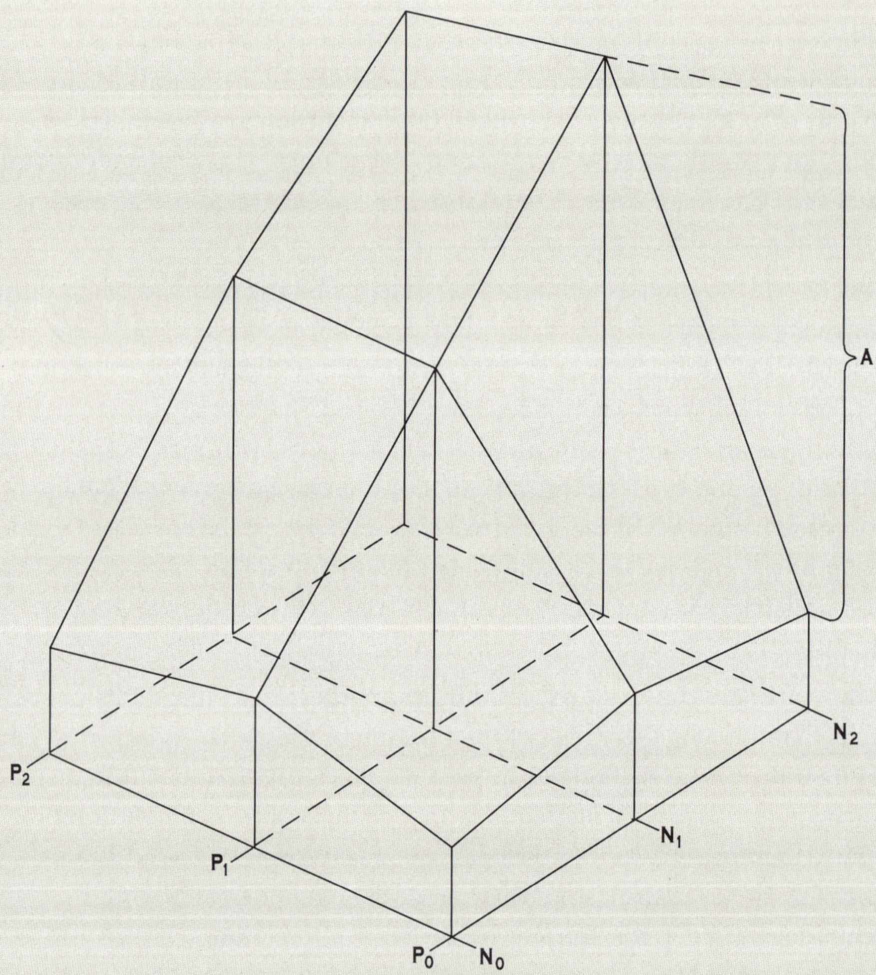

FIGURE 1

Yields of ryegrass in native pasture soil, fertilised with three levels of phosphate ($P_{0,1,2}$) and nitrogen ($N_{0,1,2}$). The increased yield 'A' is a measure of the interaction between nitrogenous and phosphatic fertiliser.

Ecosystems can be defined as communities of plants and animals (often including man) together with their physical environment. Ecosystems change from time to time, and from place to place, without exact boundaries. All plants and animals, including domestic livestock and cultivated crops, are part of ecosystems. Individuals of each species form a population, and the effect of the species in a community depends upon the size of each population. In studying ecosystems, however, there are processes like nutrient cycling, energy flow, water movement and the whole physical environment that are influenced by many species. These ecosystem processes should be defined if the behaviour of any species is to be understood. When we are interested in production of wool or meat from domestic animals, or interested in crop production, these natural processes may be enhanced or inhibited by farming practices. People's activities may have far-reaching effects and unexpected consequences. Some of these are described later.

Populations fluctuate because of changes in birthrate and mortality. Human populations fluctuate very slowly, and changes are measured in tens or hundreds of years. Mortality in infant humans is very low compared with most animals, and in domestic stock high mortality is not tolerated. It may come as a surprise that in many insect and other animal populations a massive mortality during development occurs in each generation. Also, considerable variation in fecundity in response to food supply and environmental conditions occurs in wild populations.

Table 1 shows the approximate number of eggs laid by one female of various insect species, and the number of young produced in an average lifetime by some mammals. These figures provide a simple basis for calculating mortality that will help to explain population trends. The mortalities that will maintain a stable population (all other things being equal) are listed in column a, and one wonders what could cause these deaths. In column b the mortalities that would result in a doubling of each species in one generation are given — how small these differences are! These data provide a perspective for the study of population dynamics.

TABLE 1
NUMBER OF PROGENY PRODUCED IN A LIFETIME OF EACH MATED PAIR IN SPECIES CONSIDERED PESTS

Species	No. of young (per female)	Lifespan (months)	Percentage mortality a For stable population	Percentage mortality b To double population
Eastern grey kangaroo	*12*	*100*	*83.0*	*67.0*
Rabbit	*20*	*24*	*90.0*	*80.0*
Dingo	*40*	*120*	*95.0*	*90.0*
Pasture scarab	*30*	*12–24*	*93.0*	*87.0*
Bush fly	*100*	*2*	*98.0*	*96.0*
Cabbage butterfly	*350*	*2*	*99.4*	*98.9*
Oncopera *moth*	*2000*	*12*	*99.9*	*99.8*
Field cricket	*2500*	*12*	*99.9*	*99.8*

Look especially at the mortalities of the insect pests in table 1. Think over your own experience of pasture insects like scarabs, crickets and grass grubs, for which it has not been necessary to apply insecticides on your property for 100 years. In your garden try to recall the occasions when you have seen cabbage butterflies or cutworm moths, both of which lay thousands of eggs, yet they have seldom caused damage to your flowers or vegetables. When you realise that this means there are natural mortalities each year of around 90 per cent you may wish to enquire further into nature's way of pest control.

Death strikes down insects at every turn. Nature's weapons are climate, disease, predators and parasites. The eggs are desiccated in spells of dry weather or flooded by rain or killed by disease. Predators feed on eggs and parasites lay their eggs in them. The larvae and nymphs, mostly soft-bodied, fat and juicy, are sought after by a horde of hungry predators. For each insect that feeds on plants, there are many insectivorous animals for which it is food.

PLATE 1

A simple dam with dead trees standing in water, that was to become a roosting place for 1000 straw-necked ibises, as well as a feeding ground for a few white ibises and many other water birds.

PLATE 2

Straw-necked ibises feed on insects such as crickets, grasshoppers and scarabs. Flocks of 500 and more roost at night on dead trees, especially trees over water.

PLATE 3

White ibises nesting at Healesville, Victoria. The usual breeding places are wetlands and marshes.

PLATE 4

PLATE 5

Colourful fruiting bodies of soil fungi appear infrequently in grassland, but the vegetative growth of hyphae is abundant in litter at all times.

PLATE 6

A pleasant vista of native grassland, framed between a yellow box on the left and a stringybark on the right. Early settlers were charmed and delighted by this countryside.

The adult insect too leads an uncertain life, subject to the vagaries of the weather and hunted by swifter animals, all hungry for a meal. Together these natural controls achieve a miracle of ecological balance unless disrupted by thoughtless agricultural development.

Here in a few words is what this book is all about. It is a fascinating story, all the more wonderful because you can go out into your fields and your home garden and see the phenomena described.

Predators and parasites that feed on other insects are to be found everywhere. Together they exert a powerful control on pests. You will probably recognise the species that we have photographed, mostly in their natural habitat. You will not be unmoved by the beauty of many of them, and the grotesqueness of some.

One of the weaknesses of pest control worldwide in the past has been the application of artificial control measures without consideration of biological control agents. Most pesticides are toxic not only to pests but also to predators and parasites. The worst practice is the use of insecticide as a preventive measure, before it is known whether the pest would have done serious damage. Yet it is a common practice. In these situations predator and parasite populations cannot develop, both for want of their host species and because of the effect of toxic residues.

World statistics show a remarkable anomaly. In spite of an increase in the use of insecticides, the percentage loss of crops caused by pests has increased. Jeanette Conacher in the book *Pests, predators and pesticides: some alternatives to synthetic pesticides* says that despite a ten-fold increase in the use of pesticides in the United States since World War Two, the loss of crop production to insects has approximately doubled. One reason is that pesticides are not even compensating completely for the reduction of natural controls. What a waste of money!

There is another reason why pest problems are getting worse, which is a sad part of the world's history. Agricultural development that results in complete clearing of native vegetation, and cultivation of vast acreages without leaving suitable habitats for predators and parasites, has seriously reduced natural controls. When this has gone too far it will no longer be possible to choose natural control of pests. Do **you** have a choice?

Photograph 2

The rabbit found Australia much to its liking, with abundant food most of the time and few predators. On the northern tablelands, biological control by myxomatosis and a virulent enteritis which kills many young keeps numbers down to manageable proportions. Foxes and birds of prey (eagles, kestrels, hawks, owls and ravens) also help to keep rabbit numbers down.

Photograph 3

Eastern grey kangaroos have a low reproduction rate (about 12 young in a lifetime) but their numbers increase fairly quickly because of low mortality.

One of the aims of this book is to answer this question by describing the disturbances to the eucalypt woodland ecosystem following agricultural development, and by evaluating carefully the present disturbed ecosystem in which man and domestic animals play a dominant role. What we all have to learn is that a grazed pasture is an ecosystem which cannot be defined simply in terms of grasses, clover, cattle and sheep. Counting the fungi, bacteria and minute animals crucial to processes such as nutrient cycling would reveal hundreds of different species in one paddock. Perhaps 1000 would be a closer guess if we include the birds, and the insects on which these birds feed. There have been lamentably few ecological studies of pastures. A pity, because if we had information on the population dynamics of these hundreds of species, we could form a reliable opinion about their contribution to ecosystem processes.

Pastures would not function without adequate numbers of many of the native species. Nutrients would be immobilised in undecayed organic matter if the vast array of decomposers was inadequate. A small shift in soil acidity can depress the activity of soil micro-organisms which break down soil detritus and mineralise the nitrogen and phosphorus to forms that plants can absorb. In many other ways the trampling of stock, their grazing and their excreta are slowly modifying the pastoral environment.

Our early agricultural practices, such as the ringbarking of trees and burning of the coarse native grassland, were the first impacts on a stable ecosystem of eucalypt woodland. Since then other changes more disruptive still have come about. This is not a criticism of these changes; the increased productivity of sown pastures, fertilised at great expense, has provided food for a great number of animals. Besides domestic stock these pastures provide a habitat for a steady increase in many indigenous species. Pasture insects, which previously went unnoticed, have become pests in this improved pasture. In addition, some alien species have been inadvertently introduced, and the shock of the rapid and unwelcome spread of the blue-green aphid and the spotted alfalfa aphid is still keenly felt by farmers.

The northern tablelands were settled by Europeans from 1832 onwards. The squatters in the earliest days paid an annual rental of less than a farthing per acre, and the squattages (or runs as the leasehold properties were called in early times) were very large. In 1835 there were 62 runs in the New England pastoral district, and the number of runs had increased to 171 in 1874, carrying about a million sheep, a quarter of a million cattle and 20 000 horses.

At the time of first settlement by Europeans, most of the tablelands were clothed in eucalypt woodland with native grassland beneath the trees. In many parts the trees were widely spaced, so that it was possible to gallop through on horseback or drive through the woodland in a sulky, with care. There was a sprinkling of hickory and fern-leaf wattles, slender trees half the height of the towering eucalypts. Here and there in the grassland low flowering shrubs were present — epacrids, legumes, daisy bushes, tea-trees and blackthorns — all strange to the newcomers. With a change in soil or topography the gums and boxes gave way to stringybarks and ironbarks. On some of the hills and gorges the trees became a forest, and on the Salisbury Plain the grassland was almost treeless. Only scattered trees grew on the black soil plains found in places.

In a later chapter the mood of these times is recalled, and the consequences of timber clearing considered. The early settlers cleared timber firstly by felling with axe and crosscut saw. The sucker growth and seedlings were repeatedly slashed. Later it was found that ringbarking would kill the trees, but there was a relentless task ahead of these settlers. Within 100 years the greater part of the tablelands had been partially cleared. The remaining tree density in 1962, estimated by Geoff McDonald from aerial photographs, is shown in figure 2.

Livestock numbers (see fig. 3) showed a steady increase up to 1910; by this time cattle and sheep numbers seemed to have reached the maximum that could be carried on native pastures. Most of the grasses, such as kangaroo grass and red-leg grass, were dormant in winter, while wallaby grass grew actively during spring and autumn. Snow grass or tussock was grazed in winter when little else was available except a few annual winter-growing grasses. Between 1910 and 1955 total stock numbers remained steady, with some shift in the proportion of cattle and sheep.

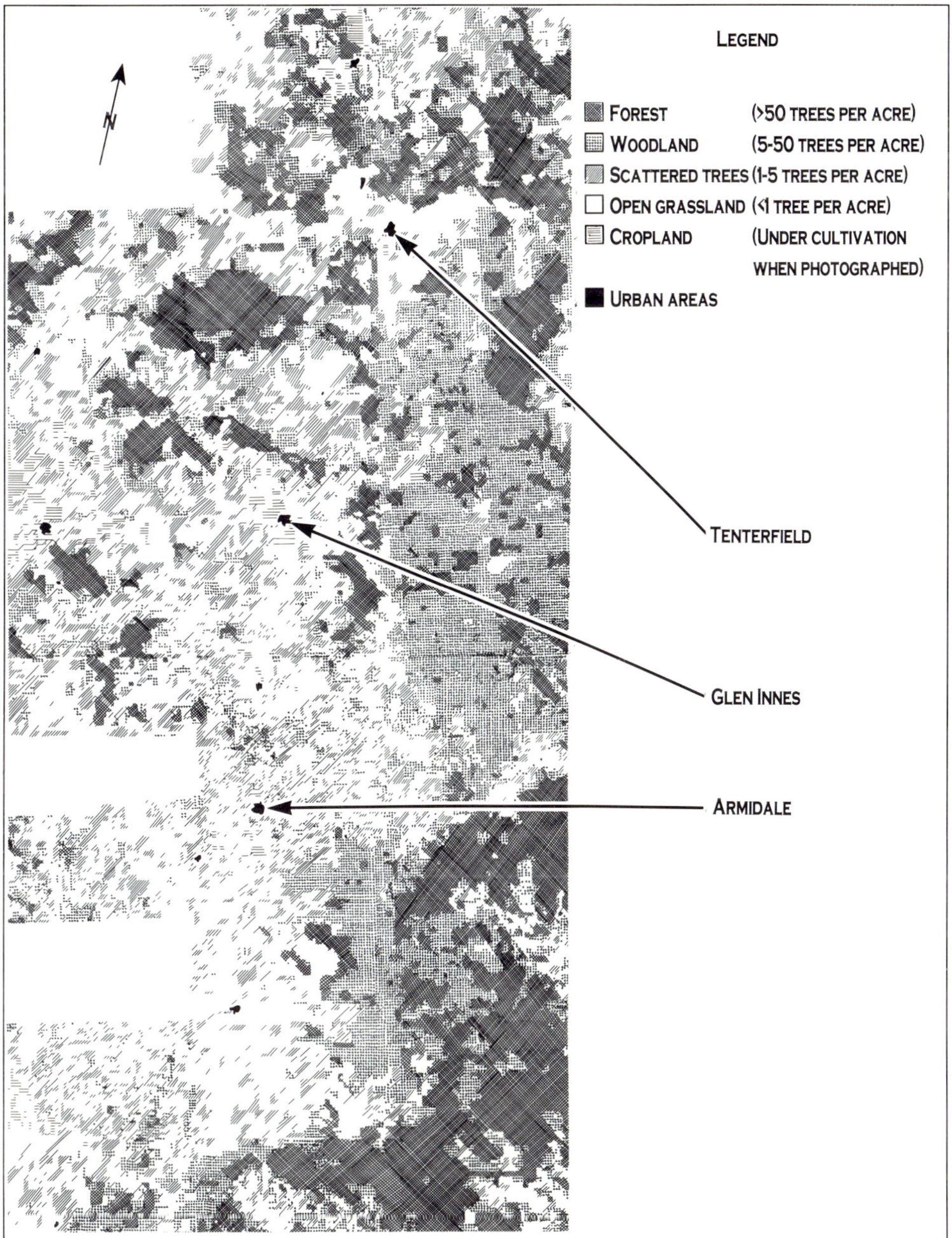

Figure 2

Map of timber density on the northern tablelands before dieback became serious. McDonald (1968) mapped these broad classes of timber density from aerial photographs taken in 1962 by the NSW Department of Lands.

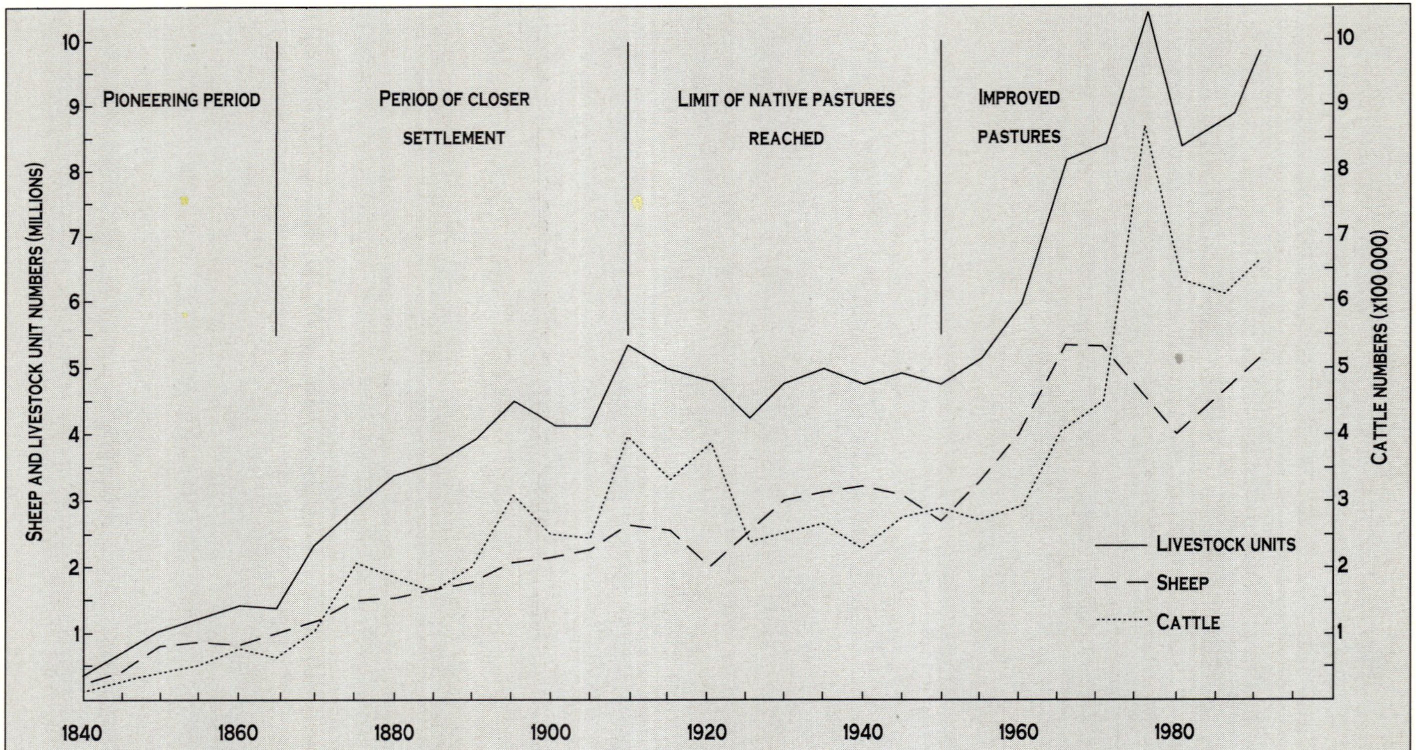

FIGURE 3

Numbers of sheep, cattle and livestock units (cattle equivalent to seven sheep) on the northern tablelands, in relation to historical development. From McDonald (1968) amended, by courtesy of the Geographical Society of New South Wales Inc. and the Australian Bureau of Statistics.

From 1950 onwards there was a change in the tablelands pastoral industry, arising from two related factors. Firstly, in the thirties graziers at Deepwater, Guyra and Walcha found that introduced grasses and clover could be successfully established if superphosphate fertiliser was applied. Unfortunately, fertiliser was unavailable during World War Two. Research later showed that there was a widespread deficiency of phosphate and nitrogen, and in some soils a deficiency of sulphur. White clover grew vigorously in this climate when superphosphate was applied, and the nitrogen fixed by bacteria in clover nodules was sufficient for perennial ryegrass, phalaris and cocksfoot to provide a mixed grass/clover sward which carried sheep and cattle throughout the year. However, the difficulty of spreading fertiliser with mechanical implements, on uneven land with scattered trees, severely impeded the development of pasture improvement.

Then in 1950, A.S. (Poss) Nivison persuaded East West Airlines to convert a small plane to apply fertiliser from the air, as he had seen planes doing in the hill country in New Zealand. Seed was broadcast in the same way on native pasture being grazed by cattle and sheep. Although the transition to a temperate grass/clover sward was gradual (requiring about four years), there was no need for cultivation and no period during which livestock needed to be withdrawn from paddocks. Since 1950 just about everything on the tablelands has changed. A new habitat — improved pasture — has been provided, and this is being fully exploited by cattle and sheep.

Vegetation species dominance has swung from subtropical grasses and sclerophyll trees and shrubs to temperate herbaceous species. The latter are now self-sown throughout the tablelands. Fescue is sometimes sown and it occurs as a dense grassland on river flats. Volunteer annual grasses are locally abundant as pioneers on heavily grazed sites, especially resting camps. Barley grass, silver grass and annual ryegrasses are winter-growing volunteer grasses on trampled sites which provide valuable winter feed for domestic and native animals.

Everyone on the land knows that improved pastures can sustain about three times as many cattle or sheep as native pastures can carry. But how many realise that the increased primary production also provides an increase of the same magnitude in energy and food for other animals — pasture insects that feed on foliage (like crickets, grasshoppers and caterpillars) and those that feed on roots (scarabs) and animal excreta (dung beetles and bush flies)? And what is the population response of insectivorous birds, lizards and spiders?

When these agricultural ecosystems are viewed as populations of many species, we realise how little we know about the stability of the systems which yield our beef, mutton, wool and cultivated crops. Stability is judged on long-term trends. In farming it is an economic assessment of two parameters, namely the value of the yield and the cost of the inputs necessary to maintain the yield. The profitability of a farming operation will vary from year to year in response to weather (which we cannot change), and it will also vary in response to fluctuations in the cost of maintaining soil fertility and controlling pest species. Predicting market values makes little sense when the biological components of the system are unpredictable. In the following chapter fluctuations in animal populations are discussed, and we shall see how these oscillations can obscure long-term trends.

3

Ups and Downs of Life

Students of wildlife soon become aware of the many dimensions of variation in animal populations. Spatial variation means simply the differences in number of individuals from place to place. This makes animal census work tedious, because no simple method of counting gives results that are representative of the whole territory. Topography, food sources, nesting sites, vegetation cover and soil type may explain some of the spatial distribution. However, some of the variation always remains unexplained. For those who estimate insect pest abundance, the differences in population density from paddock to paddock are a burden.

The population of any insect species monitored at one site reveals that the numbers go up and down. These fluctuations in abundance will decide whether a species is regarded as a pest, just a nuisance or quite harmless. One of the curl-grubs found in pastures is *Sericesthis nigrolineata*. It has the common name of dusky pasture scarab, and it is a relative of the colourful Christmas beetle. Though much smaller than most other scarabs, the grubs occasionally build up high densities in pastures, and small patches of pasture can then be rolled up like a carpet because the roots have been eaten just below the surface. The adult is a small brown beetle which flies to eucalypt trees at night to feed and mate. The number of adults caught each night in a light trap suspended in a red gum tree at Armidale throughout summer is illustrated in figure 4. The first few beetles came to the trees in November, and the last of the season were found in March. During this period of annual prevalence, however, the numbers fluctuated between 3000 on a warm still night to zero on many nights when rain, cold or wind inhibited beetle activity. Unfavourable weather prevents feeding, and under these conditions the female beetles may resorb the eggs in their ovaries.

The number of beetles caught in such traps is only a small fraction of the population present in the area being trapped. On one night 640 beetles were caught, then marked with paint and released. On the seven nights following their release, 33 marked beetles were counted in 1540 collected. A rough estimate of the population from these figures would be 30 000 adult beetles in the vicinity of the trap.

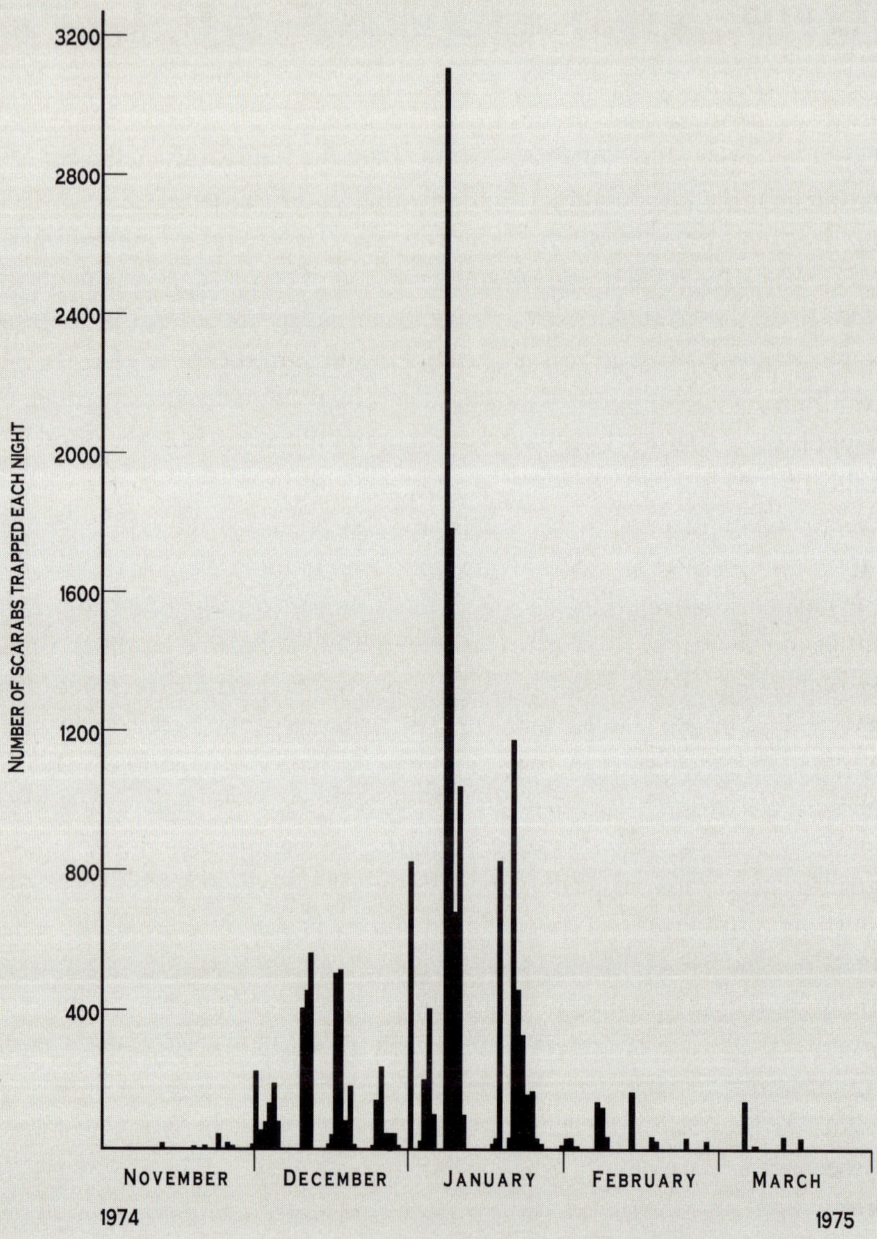

FIGURE 4

The number of dusky pasture scarab beetles trapped each night during the 1974–75 season in a red gum tree in Armidale. These beetles are prevalent in high numbers only in December and January, and during this time wind and cold may inhibit feeding on many nights.

Although the weather prevailing in the early evening affects the catches in light traps, the annual total catch may be used as a rough basis of comparison from year to year. Reg Roberts has provided data of annual catches in a trap at the CSIRO Pastoral Research Laboratory at Chiswick. Over 18 years from 1963 to 1980 the totals varied between 2000 and 17 000 (see fig. 5). It is impossible to discern any long-term trend in these figures. During the same period the number of Christmas beetles (*Anoplognathus* spp.) caught in the same light trap declined significantly. The numbers of two other species of scarabs that do not feed as adults seem to have increased slightly.

In figure 5 the progress in pasture improvement on the properties surrounding the light trap is included in the diagram. By 1910 most of the eucalypt timber would have been cleared, leaving less than one tree in ten on average. By 1990 about half the remnants had been lost. Regeneration of trees has been negligible on grazed pastures; the few remaining eucalypts are aged and most will become moribund in a few decades. All the scarab species found on the northern tablelands today are believed to have been present in the eucalypt woodland at the time of European settlement. No evidence exists to show that scarab numbers increased after clearing timber, introducing livestock and burning in winter. The meagre records of scarab numbers throughout Australia (reviewed by Margaret Cumpston in 1941) serve only to show that some species have been damaging pastures, crops like sugarcane, and eucalypts and other trees since the early 1900s. Knowing the ups and downs in scarab populations, isolated records are clearly inadequate to define the average population levels in the past. The question mark in figure 5 represents an unanswerable question.

The widespread use of superphosphate and improved pasture species (grasses and clover) may have initiated a much greater response in scarab populations than clearing and grazing native grassland. Herbage production on pastures has increased three- to four-fold. On fully improved properties livestock numbers have increased correspondingly. Excreta and plant detritus (including roots) have been added to soils, and the organic matter content of pasture soil has roughly doubled. Scarab grubs are on a low plane of nutrition in native pasture, compared with improved pasture.

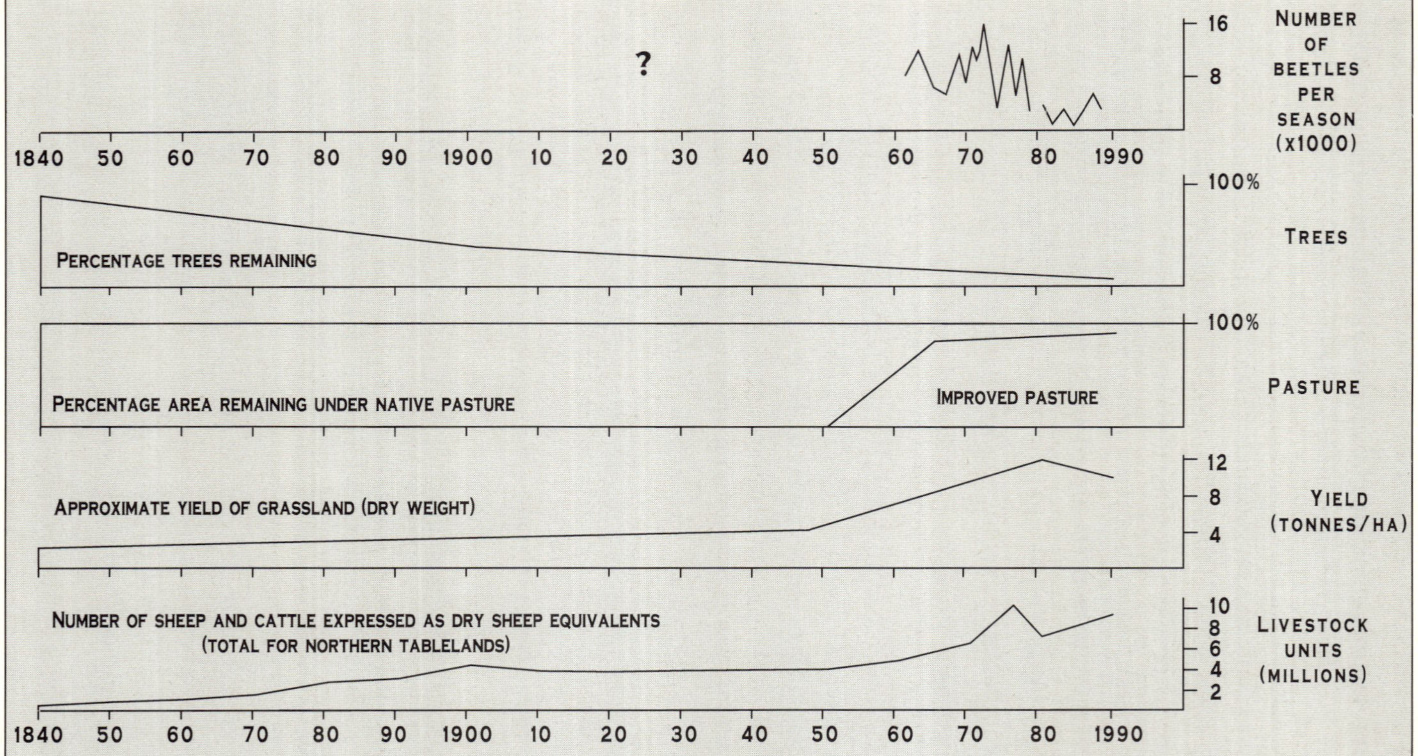

Figure 5

History of pasture improvement on 'Chiswick', the CSIRO Pastoral Research Laboratory, originally part of 'Saumarez' property. The data for the dusky pasture scarabs caught each season over 18 years at 'Chiswick', and also over seven years at 'Arding', from Reg Roberts and Gus Campbell.

Many experiments have been done to find out how fast grubs grow in different pastures. In a pot experiment young grubs of *Sericesthis nigrolineata* were kept in a glasshouse for five months in various soil, plant and fertiliser treatments. At the end of the experiment the weight of grubs in native pasture was 31 mg, while those in fertilised ryegrass weighed 65 mg. In another experiment some grubs completed their life cycle in one year under fertilised ryegrass, but took two years to complete their life cycle in native pasture soil. The higher plane of nutrition resulting from pasture improvement would also increase the fecundity of these scarab species. Despite this, however, none of the four species of *Sericesthis* showed an increase in numbers between 1963 and 1980, and the two common species of *Anoplognathus* showed a marked decline. It must be concluded that some environmental factors have prevented any increase in scarab numbers. Soil under improved pastures, for example, is drier and hotter than that under native pastures with trees (see ch. 12).

While it is interesting to speculate on the past history of scarab populations, of far greater interest are future trends in the abundance of these potential pests. Two questions need to be answered before the size of the problem and the necessary control measures can be decided. The first question is: Will numbers increase or decrease over the years to come? On the tablelands we are dealing with disturbed ecosystems in which the grassland productivity has been increased several fold, so that livestock numbers have doubled in 25 years. Some indigenous animals have dwindled and some new domestic species have become naturalised. We have mentioned that the improved food supply for scarab grubs produces bigger female beetles that will lay more eggs. However, there have been other changes. In places eucalypt trees, on which the scarab beetles feed, have decreased to about a tenth their original number. The physical environment has been greatly altered, and in the next chapter we discuss the effect of this upon mortality in scarab populations. The many factors which contribute to a shift in insect abundance can be clearly defined experimentally, and then put together. That is the only way to answer this first question.

The second question is: How will the population fluctuate from year to year? In many insect populations these wide fluctuations can be explained by vagaries of the weather, which affect fecundity and mortality. The chances of survival can be predicted when the precise causes of mortality are identified. In chapter 4 the fascinating topic of environmental stress is explored, and we will see how much of the population dynamics is explained by prevailing weather. The rest of the book deals with mortality caused by predators, parasites and disease. Here we should dwell for a moment on the significance of the year to year fluctuation in pest abundance, and consider its economic implications.

All crops and pastures can tolerate a surprisingly large amount of feeding by pests before the yield is reduced. There is no reason to be surprised about this. Orchardists prune fruit trees to enhance yield; gardeners disbud ornamentals to obtain large flowers; vegetable gardeners thin out seedlings of all kinds of plants; and we mow our lawns to maintain the vigour of the grasses and clover. Graziers know that adequate stocking intensity (about eight sheep to a hectare throughout the year) sustains the vigour of pastures. More lightly grazed pasture accumulates moribund foliage. Should we not expect the same in roots of pasture species?

Having studied pasture growth from many angles — intensity of stocking with cattle and sheep, response to fertilisers and damage by insects — we can answer the question of tolerance to defoliation and root-feeding with quiet assurance. Plants have the capacity to replace leaves or roots when these are damaged; regrowth is rapid and in a few days foliage and root functioning are balanced again. Stem bases, leaves and roots contain a pool of carbohydrate and other substances, which may be called the regrowth potential. Damage is often repaired without any interruption to normal growth rate. If moribund tissue is eaten off leaves that are shaded or roots that are inefficient, the plant may even increase its overall growth rate. Many years ago Irene Stuckey found that most perennial grasses replaced a large part of their root system each year.

Root growth declines as spring advances, and the carbohydrate content of roots drops lower and lower. By summer root growth is very slow, because high soil temperatures allow efficient functioning of the root system. In the autumn, however, declining soil temperatures tend to inhibit absorption of minerals and water by the roots, and plants compensate by an increase in the rate of root growth. While the growth of ryegrass and clover pastures seems to stop in winter, the products of photosynthesis are mostly being translocated into the roots, and new roots are being formed. In subtropical grasses, such as the native kangaroo grass and red-leg grass, a similar seasonal pattern of root growth occurs in autumn, but when soil temperature falls below 5°C roots no longer function adequately, and the leaves wither.

With this view of pasture growth in mind, let us consider the feeding of pasture insects. Experiments with scarabs in pots of ryegrass and clover in the glasshouse have shown the following:

- When scarab grubs are present the ratio of root to foliage weight is lower than when grubs are not present. This suggests that the young roots which grow after grubs have eaten off the older roots are more efficient.
- When scarab grubs feed on ryegrass or clover roots, often no damage occurs to foliage or there is only slight reduction of foliage growth unless feeding continues for many months.
- If a defoliation treatment (simulating grazing) is included the feeding on roots causes a significant reduction of foliage growth. In contrast, however, in experiments reported by James Ridsdill Smith, ryegrass which was frequently defoliated was damaged less by scarab grubs, because plants were small and shallow-rooted, as explained in the next point.
- Plants which are growing well because of adequate fertiliser and soil moisture suffer greater damage than plants under unfavourable conditions. Grubs tend to feed by moving horizontally at a chosen depth, cutting off the roots at this depth, and large plants have more roots cut off than shallower-rooted small plants.
- The depth at which grubs feed greatly influences the damage to pastures. Root density and moisture and temperature gradients determine the depth at which grubs are found.
- When the damage done by various densities of grubs is compared, the curves of root damage and foliage reduction are asymptotic. That is, the more larvae present, the less roots are eaten by each larva. Very high densities of larvae in the soil cause no more damage than lower densities, a fact for which we should be very grateful (see fig. 6).

- Most surprising of all, plants are rarely killed by scarab grubs, even at densities of 12 grubs per pot, equivalent to more than 1000 grubs per square metre of pasture. Using more realistic densities, plants were not killed after four months, with 170 grubs per square metre, although foliage yield was reduced to one-third of the controls, comparing the accumulated foliage from four-weekly clipping.

Results of glasshouse experiments cannot be extrapolated directly to the field. But they do help us to understand what can be observed in pastures, knowing something about the complex processes involved in pasture damage by scarabs. Pasture growth may be reduced by scarab grubs feeding on the roots, and under certain weather conditions the reduction of carrying capacity may be as much as a sheep per hectare. This will occur only when high scarab densities coincide with periods of full utilisation of pasture, for example in dry summer spells or in early spring when pasture growth is slow.

At most times pasture herbage available to stock is in excess of immediate demand. Some loss, provided the vigour of pasture plants returns before critical times, is tolerable. Scarab feeding is a seasonal phenomenon because the activity declines in cold weather — scarabs are not warm-blooded like mammals. Scarabs are seasonal too because the grubs pupate in October, after they stop feeding about September in the tablelands. Although there may be a carry-over of younger grubs, the numbers will always decline in spring. Eggs of the next generation will not hatch until late December, and the tiny grubs will not be harmful until autumn. We can draw a comparison between the feed requirements of domestic stock and scarab grubs in relation to the growth rate of pastures.

Before defining the economic thresholds of pasture pests, one point should be emphasised. All pastures in the tablelands always have scarab grubs. This is not a cause for concern. There will usually be grubs of several species of scarab present, all sorts and sizes. In reasonable numbers they are performing an essential function of fragmenting roots, which assists in the cycling of nutrients. It would be disastrous if all scarabs were exterminated from pastures and moribund roots accumulated. We should learn to live with scarabs, and ask only that we understand how they might behave in all circumstances.

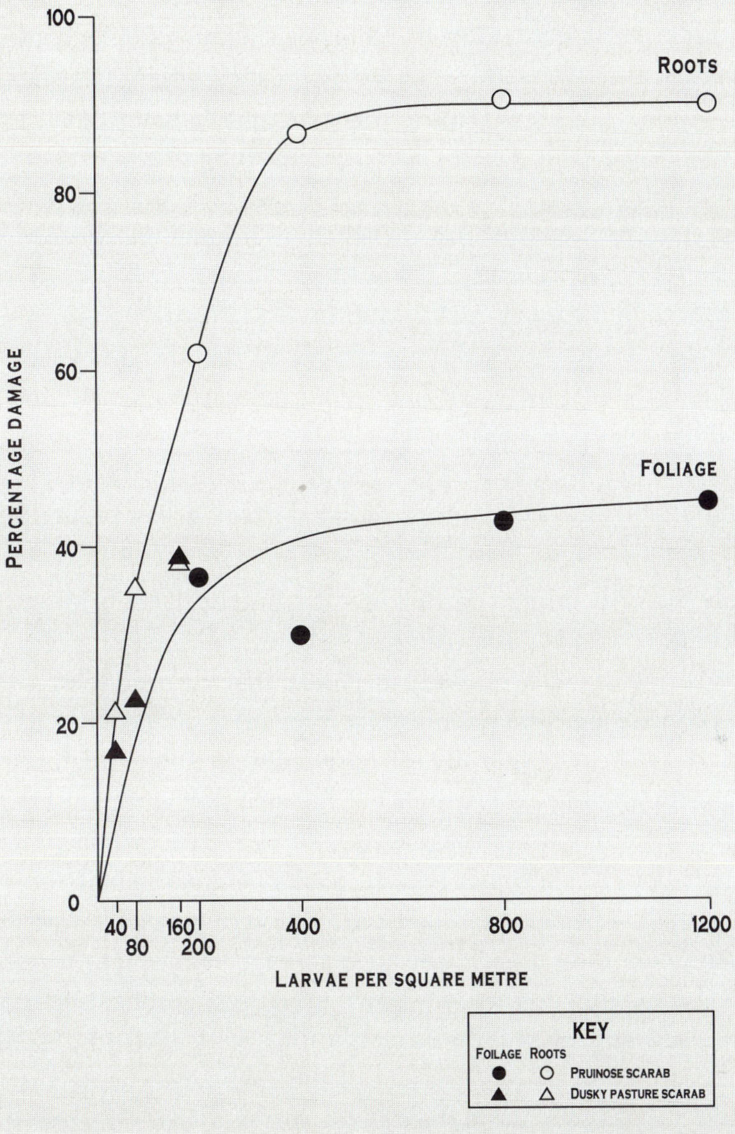

FIGURE 6

The effects of various densities of scarab larvae on root damage and foliage reduction (expressed as percentage of controls without larvae). Dusky pasture scarab data from Ridsdill Smith (1977) and pruinose scarab data from Davidson, Wensler and Wolfe (1970). By courtesy of the *Journal of Applied Ecology* and the Standing Committee on Agriculture, Australia.

Commonsense suggests that in pest management an economic threshold is reached when pest numbers reach the level at which insecticide or other controls should be considered. However, the decision on whether controls should be applied, and when and what would be appropriate, depends on many considerations. The likely damage should be predicted in its cash value, and the cost of insecticide applications or other farming practices estimated. The cost–benefit ratio should be favourable. Furthermore, any side-effects such as an adverse effect on predators or other environmental consequences should be taken into the accounting. For example, there are many cases where insecticides, especially broad spectrum chlorinated hydrocarbons and organophosphates, reduced predators so effectively that the pest (after an initial decline in numbers) increased to a higher population density than before treatment. Other pest species may also increase after insecticide application. Where this is likely to occur, the decision on the use of control measures should take into account the predicted frequency of pest outbreaks in years to come.

Let us apply the principles to pasture scarabs. Firstly, the only appropriate measure of economic loss is the predicted loss of animal products, or a reduction in carrying capacity of pastures. Notice that we need to **predict** the loss, because a control measure is required to **prevent** loss. Secondly, pastures are seldom grazed so heavily that all herbage is utilised. Keith Hutchinson has found in long-term grazing experiments at Chiswick that with ten sheep per hectare only 58 per cent of the herbage is utilised by the sheep on average. At a stocking intensity of 20 sheep per hectare, 85 per cent of the herbage is utilised, and a small reduction of foliage production caused by scarab grubs could reduce animal production. Using such average values, four thresholds of damage can be roughly defined, as shown in figure 7. A most important point is emphasised here: even at the economic threshold, where the estimated loss of animal production exceeds the cost of control by insecticide, the decision should not be based upon the cost–benefit ratio in a single season. In the diagram, sheep production would only be affected in three years out of 18 years, and spread over 18 years the loss is negligible. Even if the number of grubs exceeds 400 per square metre (which occasionally happens in localised patches), and it is predicted that carrying capacity will decline by several sheep per hectare over the spring and early summer, the use of insecticide

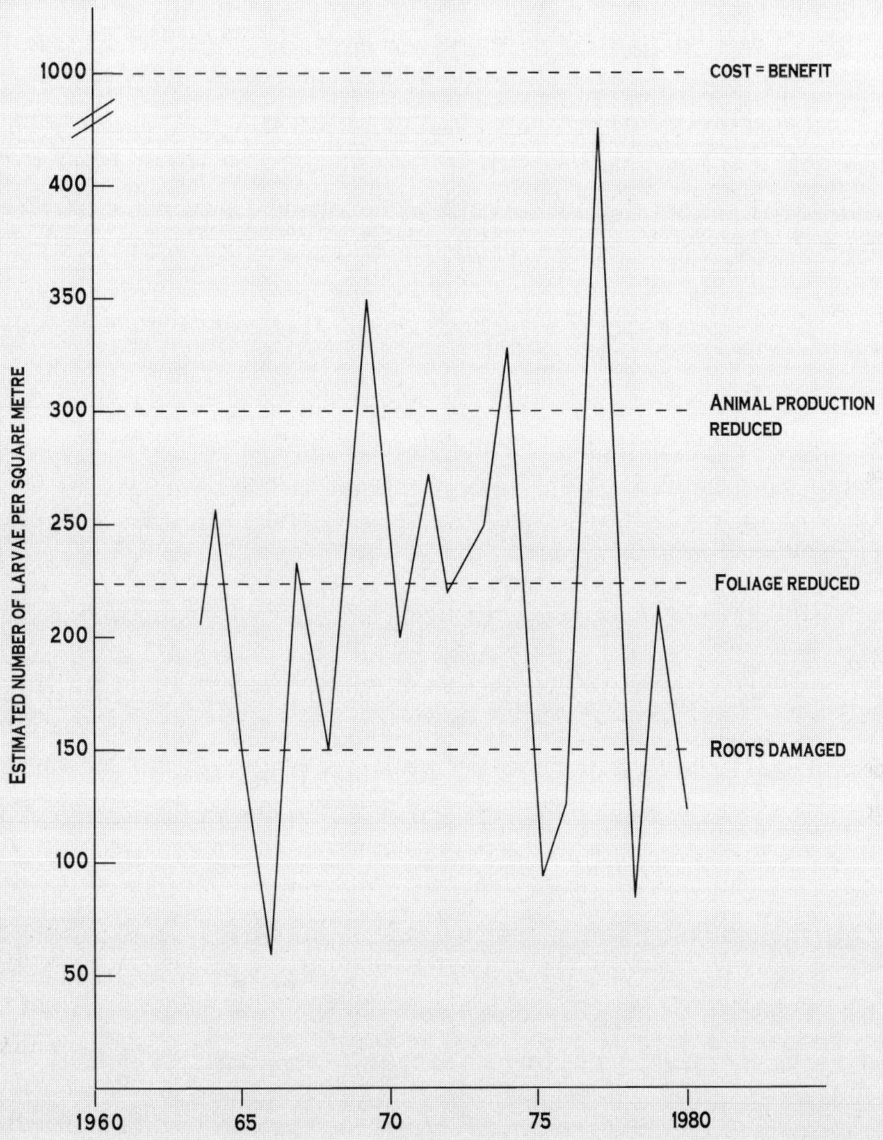

FIGURE 7

Estimated numbers of pasture scarab larvae in spring at 'Chiswick' 1963–80 and the approximate thresholds for root damage, foliage reduced, animal production reduced and animal production losses equal to cost of insecticide treatment. Actual thresholds will depend on species and pasture growth. From Davidson (1982), by courtesy of CSIRO Australia.

against soil animals like scarabs and root weevils is not advocated. Scarab outbreaks are sporadic, as shown in figure 7, and dense aggregations of their grubs are localised in pastures. If only one-fifth of a grazing property is severely affected, three outbreaks in 18 years overall means that any one paddock will suffer damage just one year in 30 years.

To kill these subterranean grubs, sufficient insecticide is necessary to make the soil toxic. If organochlorines or organophosphates are used at effective dosages, all predators in the soil (carabids, elaterids, asilids, spiders and parasitic wasps) would be killed. With some treatments, insectivorous birds would also be killed if they feed on dead larvae. Therefore, to the cost of the insecticide treatment should be added the predictable consequence of increased pest numbers in years to come. At present on the tablelands, the low frequency of scarab damage is in some measure dependent upon natural control by predators. We should be thankful that we still have thriving predator populations in our pastures, and learn how to ensure their continued survival.

In Australia, wool and meat are produced from animals which graze year-long on perennial pastures. For the most part the flocks of sheep and herds of cattle are set-stocked, each property carrying the number of head that will utilise efficiently the herbage produced in an average year. In years of high rainfall there is an over-abundance of feed, and in these years the ravages of scarabs, *Oncopera* caterpillars, crickets and grasshoppers are inconsequential. No farmer of sound mind would spend good money on control measures, unless he or she feared that leaving them would allow the species to build up even bigger populations in the years to come. Fortunately, farmers may rest assured that Australian pasture pests do not increase in this way. As in many other pest species, the high densities favour natural controls.

On the northern tablelands there is usually an abundance of herbage available during the seasons in which most pest outbreaks occur. Grasshoppers and crickets are most numerous in summer, and this coincides with the summer flush of growth of native pastures on the tablelands, except in drought years. The seasonal pattern of pasture growth in clover, ryegrass and phalaris pastures shows a less pronounced difference between summer and winter. Pasture growth rate may be slightly depressed in midsummer.

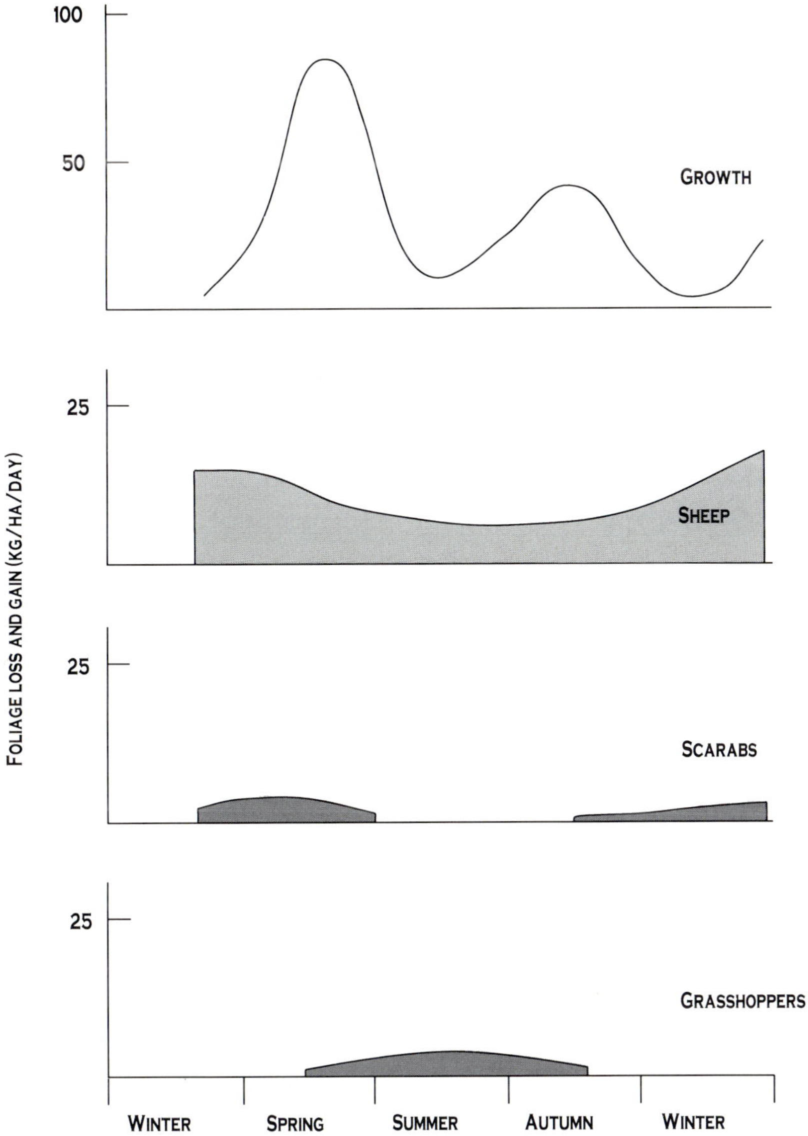

Figure 8

Rough estimates, expressed as dry weight, of: growth rate of an improved pasture on the tablelands from data for Phalaris/clover pasture (Hilder 1963); daily intake of 10 Merino sheep; daily loss of foliage production caused through root-feeding by high densities of scarab larvae; and daily loss of pasture eaten by crickets and grasshoppers in exceptional years. From Davidson (1982), by courtesy of CSIRO Australia.

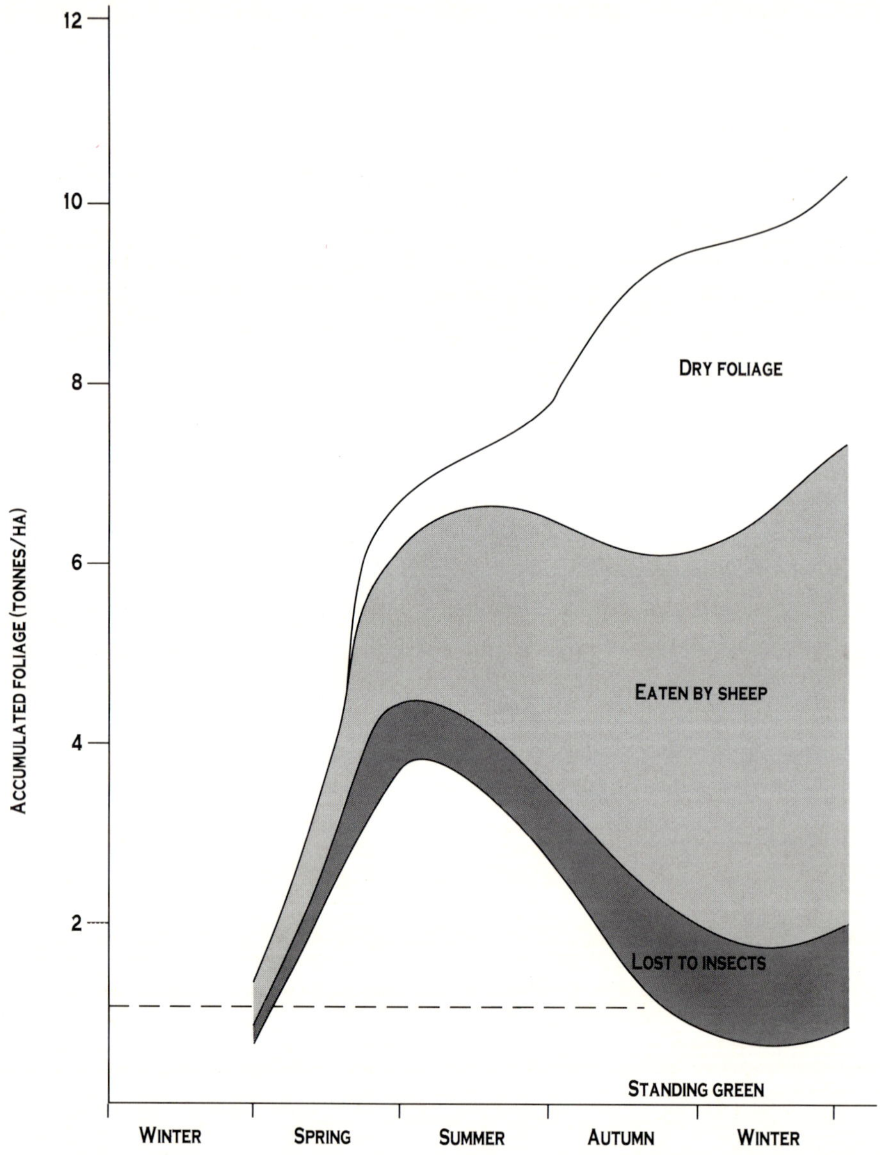

FIGURE 9

Cumulative pasture yield partitioned into dead foliage, foliage eaten by sheep and foliage lost to pasture insects (see fig. 8). All expressed as tonnes per ha dry weight. From Davidson (1982), by courtesy of CSIRO Australia.

In autumn the rainfall is more effective because evaporation is lower than in summer. Jack Hilder has described the growth pattern of phalaris/clover pasture and this is shown in figure 8, together with the estimated daily intake of ten wethers per hectare, all expressed as dry matter. In the same figure there is a rough estimate of the reduction of foliage that might be caused by grasshoppers and crickets (in summer) and the estimated reduction of foliage resulting from a heavy infestation of scarab grubs. In figure 9 these quantities are added and subtracted to give a generalised diagram of the accumulated herbage on a grazed pasture. The amount of green pasture available after all consumers have grazed or nibbled is the quantity shown below the lowest line in the diagram. Data from painstaking measurements by Keith Hutchinson and Kathy King at Chiswick are on a par in a good season (see fig. 10).

A shortage of feed is most likely to occur in late winter and spring, and in figure 10a the amount of green pasture available is below 1200 kg per ha even in a year of average rainfall. Scarabs are feeding voraciously in spring, prior to pupating. This is one time of the year when scarabs may reduce animal production on the tablelands. When available pasture falls below 1200 kg per ha, each sheep's intake is reduced by shortage of feed. In a year of below average rainfall, for example during summer drought, the pasture available may be below 1200 kg per ha for months at a time (see fig. 10B).

If stocking intensities greater than ten sheep per hectare are practised, the likelihood of seasonal feed shortage is increased. When graziers see the quantities of pasture which dry off during summer they may be tempted to increase stock numbers, but the high cost of supplementary feeding in the lean months and drought years is usually a deterrent. The variation of stock numbers to accommodate the seasonal peaks and troughs of pasture growth is seldom worthwhile in the long run. On the tablelands generally a stocking intensity above ten dry sheep equivalents per hectare is not sustainable, and in the long term it causes pasture deterioration and soil degradation.

Pasture insects should be regarded as a hazard no worse than cold, drought and falling prices for marketable products. Decisions about management practices should be based on long-term averages and probabilities.

Each species, whether it be rabbits, scarabs, cotton bollworms, codling moths or cockroaches, has its own combination of ecological factors which

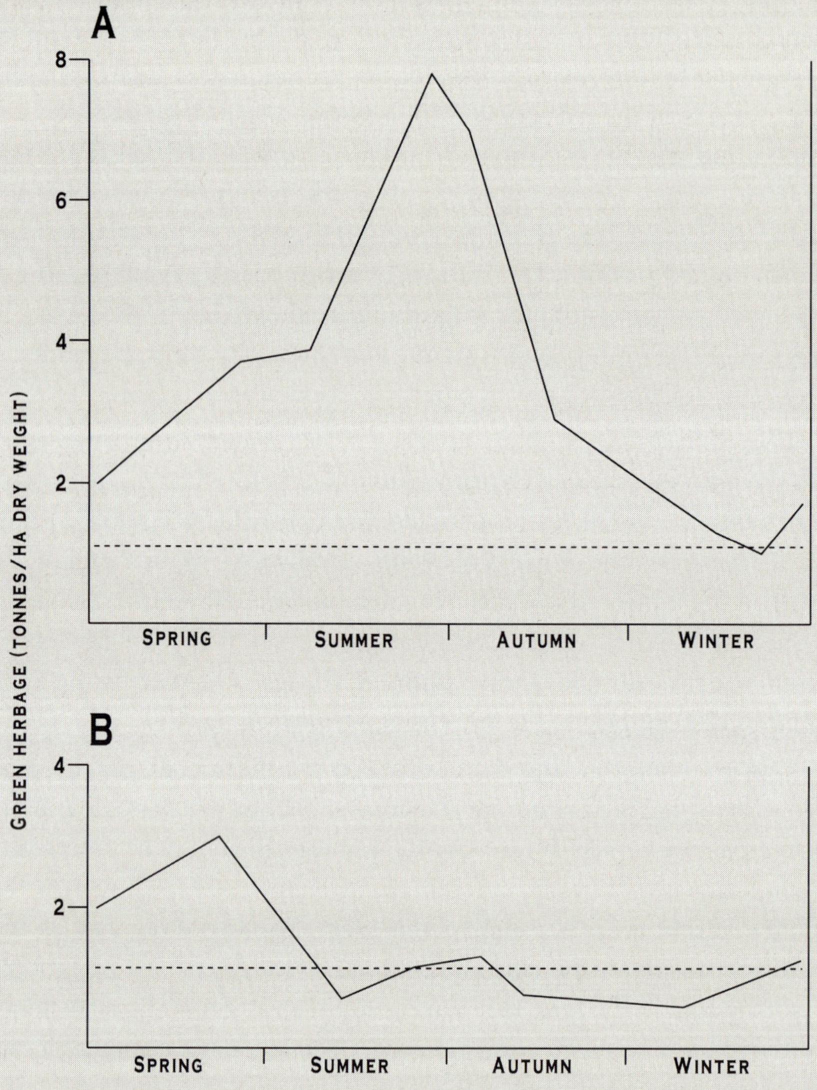

Figure 10

Dry weight of standing green herbage in a pasture grazed by 10 sheep per ha for 1971–72 (A) and 1967–68 (B), a year of unusually low rainfall. The broken line on 1200 kg per ha represents the level below which intake of sheep is reduced by the low availability of pasture.
From Keith Hutchinson and Kathy King.

determine where it occurs and the fluctuations of the population. Without an understanding of the long-term fluctuations and any upward or downward trend in numbers, no reliable decisions can be made about control. Figures 11B and 12 show the frequency of minor and major plagues of the Australian plague locust and outbreaks of wingless grasshoppers. Fluctuations of two scarab species are shown in figure 22.

The plague locust population increases rapidly in the arid interior when rainfall is above average and desert plants are green. They spread to the agricultural belt of eastern Australia and breed there if rainfall is adequate. Every plague is terminated by a dry summer. The frequency of major outbreaks in the eastern States is six in 56 years, with smaller outbreaks about every two years. In the 1984 plague (the worst on record) aerial and ground spraying were successful in preventing major crop damage, but did not affect the course of the plague as those locusts which survived the control measures died soon afterwards in the summer drought. Parasitism is not a major cause of population decline in any of the plagues studied. Spraying is necessary because few of the parasites which attack all grasshoppers and locusts extend into the western half of New South Wales (Graeme Baker 1983; Diana Wright 1987). In the agricultural belt it is rain in the usually dry summer which appears to be the critical event in the development or maintenance of a local plague. However, in north-eastern New South Wales and adjacent Queensland, where summer rainfall is usual, locust outbreaks are infrequent. In these areas the local parasite complex probably keeps locust numbers in check.

In Western Australia plague locusts are present in low numbers in the arid pastoral regions where they cause little damage. When good rains fall, locust numbers increase and they move into the agricultural regions to the south (fig. 11A). Major plagues occur once every eight or ten years. In 1982 the plague ended suddenly after an extensive control campaign by the Agriculture Protection Board (APB) of Western Australia, combined with a very hot dry summer.

The advice of the APB is to treat hoppers before they reach the adult stage. Hoppers have a number of natural enemies such as birds, spiders and wasps, but except for a small scelionid wasp, none of these natural enemies is able to

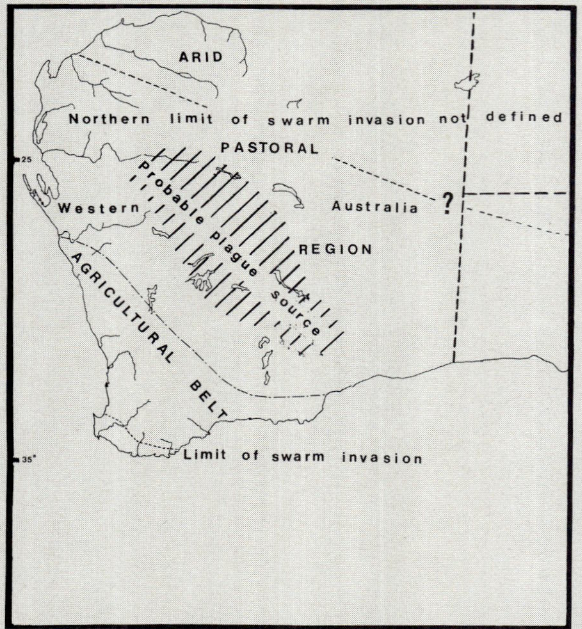

FIGURE 11A (LEFT)
In Western Australia the probable source of plague locusts is in the arid pastoral zone. The relationship between the western and eastern locust populations is uncertain. From Walden (1990), by courtesy of the Western Australian Department of Agriculture.

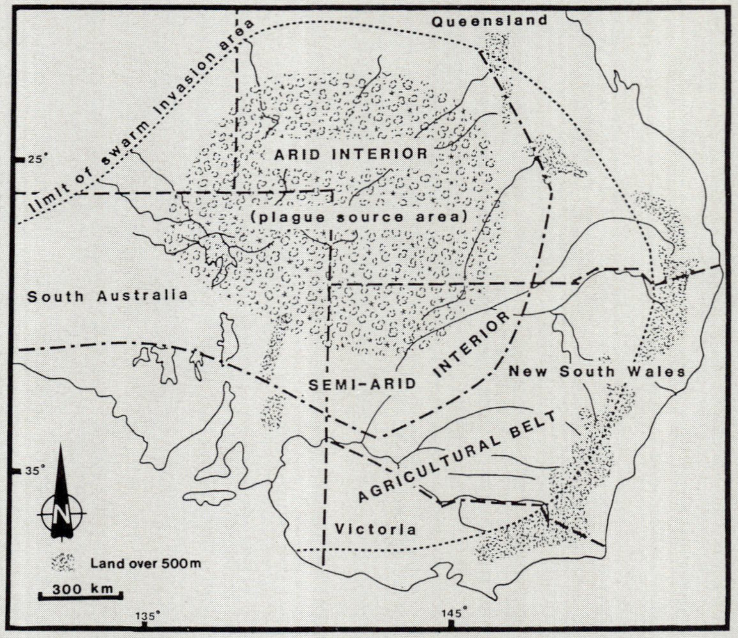

FIGURE 11A (RIGHT)
Distribution area of the Australian plague locust in the eastern States, showing the source of most plagues in the arid interior. Of the six major plagues between 1933 and 1989, only two originated in the agricultural belt. From Wright (1987), by courtesy of the *Australian Journal of Ecology* and the Australian Plague Locust Commission.

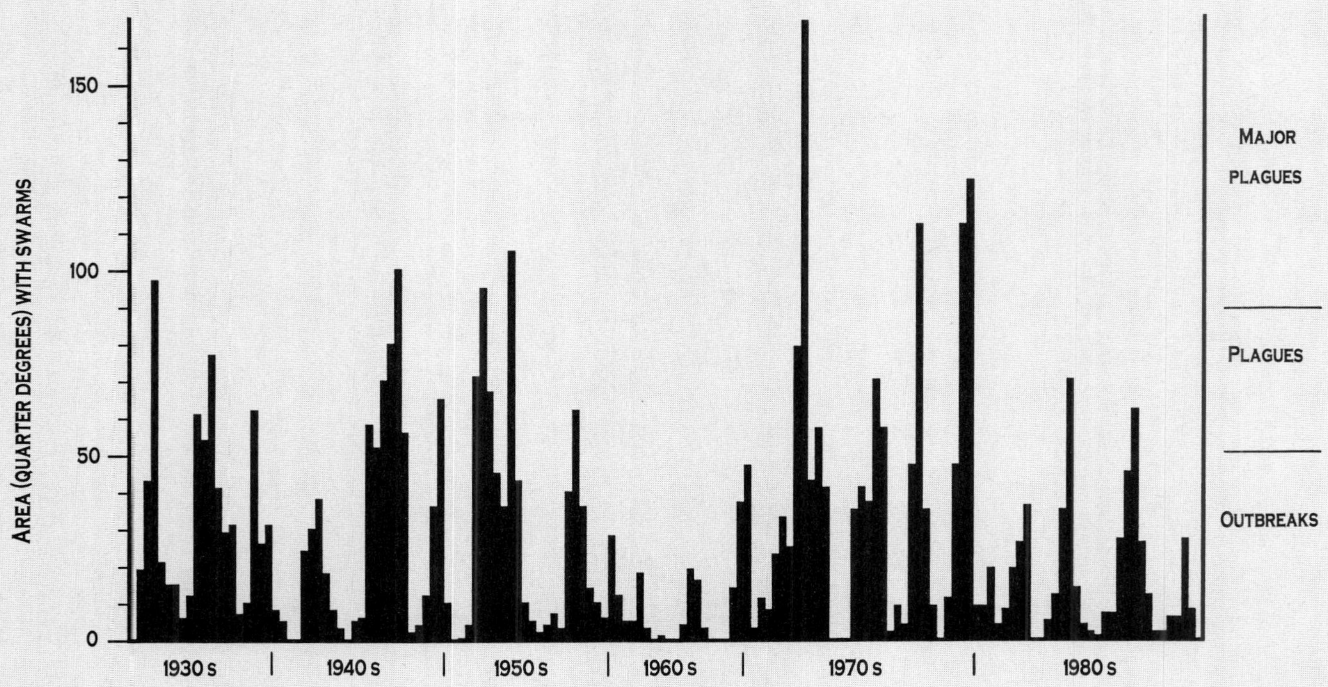

Figure 11B

Number of quarter degree squares ($\frac{1}{2}$ degree latitude x $\frac{1}{2}$ degree longitude) in which one or more swarms of plague locusts were found, in the eastern States, each locust generation 1933–89. There are two to four generations per season.

By courtesy of the Australian Plague Locust Commission.

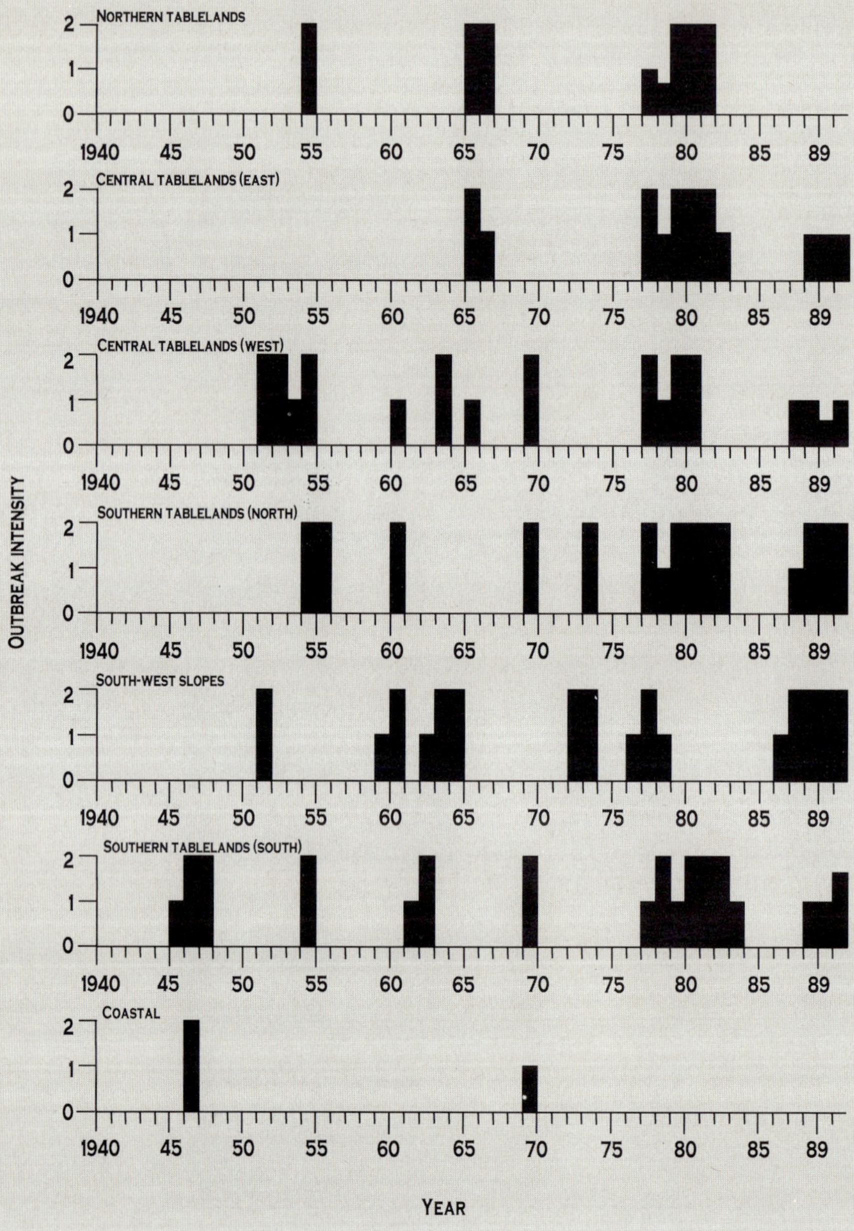

Figure 12

The seasonal abundance of wingless grasshoppers in New South Wales 1940–89. From Baker (in press), by courtesy of CSIRO Australia.

regulate locust numbers during an outbreak. The wasp parasitises locust eggs and levels of parasitism above 90 per cent have been found at some egg bed sites. Although widespread in the eastern States, scelionid wasps are recorded only from the extreme south-western corner of Western Australia (Kevin Walden 1990). Recognition of the important role played by scelionid wasps in the natural control of locusts has led to control strategies which cause minimal harm to wasp populations. Locust hatchings are not treated until they have marched away from egg beds, and the aerial treatment of locust swarms ceases once egg laying has started (Graeme Baker 1983).

The wingless grasshopper is a serious pest on the tablelands of southern Queensland, New South Wales and mid-western Victoria. Isolated pockets occur in Tasmania, South Australia and Western Australia. It is always present in low numbers in these regions, and severe outbreaks are a result of high survival of young hoppers during very dry springs, especially where intensive grazing has occurred (fig. 12). Open bare areas, especially on warm north-facing ridges, are favoured egg laying sites. Good grazing management should avoid overgrazing of these sites. Roger Farrow and Graeme Baker (in press) say that the intermittent outbreaks of this pest could be much reduced by planting perennial grasses (without clover or weeds) and trees in the source areas.

Outbreaks seem to have increased with increases in improved pastures, clear felling and heavy grazing. When the eggs hatch, strategic spraying only of these limited areas is recommended. Spraying should not be done after the hoppers disperse, that is, not after early December in the northern tablelands and mid-December in the central and southern tablelands. Natural predators and parasites of the grasshoppers bring these outbreaks to an end, and broadacre spraying is not advocated (Graeme Baker 1981). There are many other grasshoppers occurring together, including two wingless species and several winged species. All the potential pests in pastures, including scarabs, could become more serious if widespread insecticidal control were practised against any one species — with disastrous economic consequences.

Similarly, having examined the ups and downs of scarabs on the northern tablelands, we confidently recommend that natural controls are adequate to keep numbers below the economic threshold of damage. The occasional economic loss, spread over the intervening years, is tolerable. As a result of long-term research, to which many scientists have contributed, it has been possible to live with scarabs without using insecticides at all. The savings are enormous, and toxic residues in beef and mutton have been avoided.

Graham Goodyer, in an extension pamphlet in 1985, explained that control by insecticides of root-feeding scarabs in pastures is not economic on the northern tablelands because these chemicals would also be toxic to predators and parasites. The recommended method of control is to maximise the effectiveness of natural control agents by providing bushland on farms, fenced to exclude livestock, with nectar-bearing shrubs for parasites and trees where birds can nest. Renovating pastures with a fine-tooth harrow exposes the grubs to predators. Close grazing in hot weather exposes the grubs and eggs to desiccation and high temperatures, but unfortunately encourages wingless grasshoppers. Damage by pasture scarabs on the tablelands is sporadic and localised, and outbreaks occur infrequently. Unfavourable weather, predators, parasites and disease together keep them down to economically acceptable levels. Livestock production is seldom affected, because scarabs usually decline in drought years.

Whether we can be sure that a favourable balance will always continue is another matter. In later chapters we look at the role of predators, parasites, and disease in keeping insect numbers down, so that this balance can be maintained, but first we should consider the mortality caused in insect populations by extremes of the weather.

4

CHANCES OF SURVIVAL

One of the most intriguing stories in biology concerns the impact of environmental stress on animal populations. If you stride over pastures, ride on horseback, or use vehicles to round up stock, you do not see in correct perspective the environment of small animals. To insects in pastures the shadow cast by a tussock of grass can be the difference between life and death from a merciless sun. Each hoofprint becomes a lake in the rain. For those insects which burrow in soil the environment is a tunnel which fills with water after each heavy downpour.

Some years ago we set out to search for the causes of pasture insect mortality. We were determined to see the soil environment from the insect's point of view. Temperatures, moisture and the gases in the soil spaces were measured with minute, sensitive instruments.

Frankly, we were aghast at the excessively high temperature reached in the soil in the temperate climate of the tablelands. On a hot day in summer the topsoil under pasture heated up above 42°C in heavily grazed patches. Most remarkable were the gradients between the sunny side of a plant and the shaded side, and the fall off in temperature with depth. Moving around in a cubic foot of soil under pasture, a grub could be exposed to a climate difference comparable in human terms to the climates of Darwin and Canberra, all in the space of a hand's span (see fig. 13). Between sunrise and early afternoon a newly-hatched larva of any scarab (some only 2 mm long) would have to move fast against the rising temperature to avoid being killed. This occurs on several days each summer. Many larvae do not escape, and mortalities up to 90 per cent are recorded in very hot summers. A few hot days in a year spell a quick death to many insect pests in the soil.

Death from heat stress was recently shown to be extremely important in scarab population dynamics. From careful studies of heat tolerance in several scarab species, the threshold for stress was defined between 32.5°C and 35°C, and for the black-soil scarab found on the western slopes it was slightly higher. Near these threshold temperatures, exposures of six hours caused only low

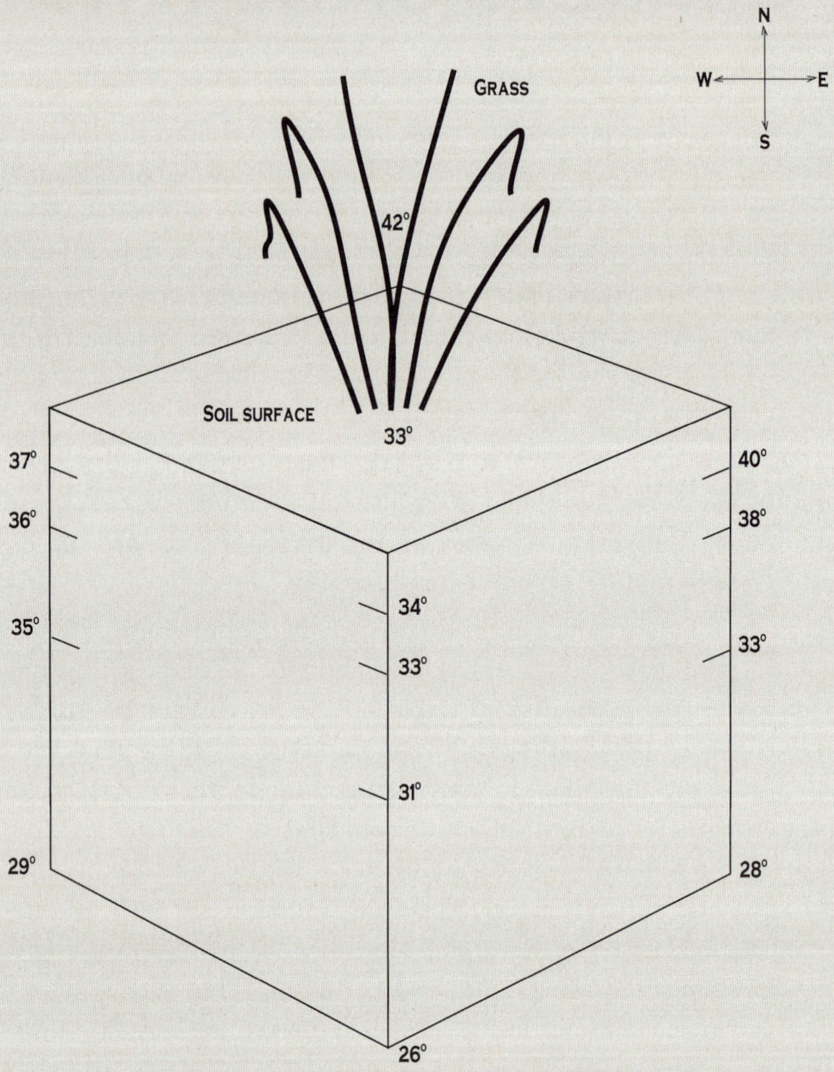

Figure 13

Soil temperatures in a space 20 cm x 20 cm with a tussock of ryegrass in the centre; temperatures measured in °C at 2.5, 5, 10 and 20 cm below the soil surface.

Plate 7

Aerial application of superphosphate, a familiar sight on the northern tablelands — the second great impact of man on this ecosystem. Photo by David Boundy, courtesy Superair, Armidale.

Plate 8

The aim of pasture improvement was a green grass/clover pasture, with scattered shade trees and stock watering places.

PLATE 9

A bee fly which lays eggs in knot-holes in fence posts and dead trees; the adult fly does not feed. Photo by Max Hill, courtesy NSW Agriculture.

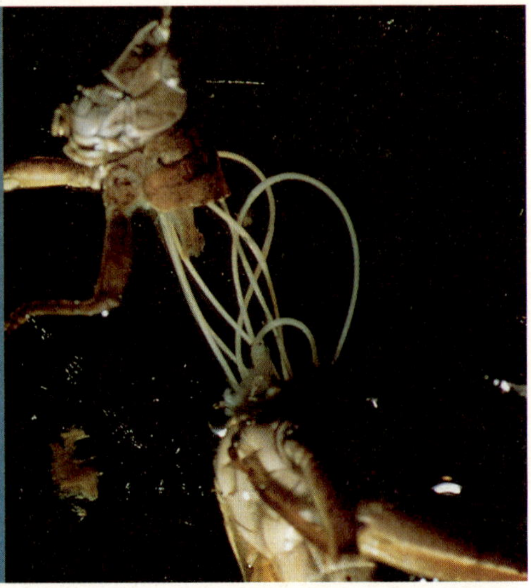

PLATE 10

A nematode worm emerging from a wingless grasshopper; the worm will lay thousands of eggs in a cell deep in pasture soil. After rain in spring the tiny larvae wriggle to the surface and infect grasshoppers. Photo by Max Hill, courtesy NSW Agriculture.

PLATE 11

A hover fly maggot holds an aphid aloft before devouring it.

PLATE 12

Adult hover flies sip nectar and honeydew to obtain energy for reproduction, but the maggots feed on aphids.

Plate 13
Harlequin ladybirds, both as larvae and adults, consume huge numbers of aphids.

Plate 14
A robber fly watches for airborne prey.

Plate 15
Paper wasps feed their larvae every day on chewed-up insects, and build up the cells as the larvae grow. Eggs can be seen in the upper cells.

PLATE 16
Courting lacewings. Their beauty belies their predatory disposition.

PLATE 17
Ground beetles are fast-running insects with strong mandibles for seizing the prey they find under litter or rocks.

PLATE 18
Click beetle larvae, called wire-worms, may be either herbivorous or carnivorous, but both groups inhabit the soil.

PLATE 19
A mating pair of a Western Australian thynnid or flower wasp (*Elidothynnus* sp.) feeding on red-flowering gum. Photo by Babs and Bert Wells.

percentage mortalities, but with higher temperatures exposure of an hour or less caused high mortalities. At 40°C (for comparison, this is about the body temperature of cattle and sheep) most scarabs are killed outright in about an hour. Christmas beetle larvae, which are slightly more tolerant of high temperature, showed a very low survival rate after three hours at 40°C. Even a brief rise in soil temperature to 42°C is lethal to most soil insects.

Having learned to recognise the precise conditions in which heat stress occurs, we can predict the deaths that will occur in these pasture insects in the spells of hot weather in summer. In years in which summer is cool because the sky is overcast with persistent rain, egg hatching and survival of young scarab larvae is good, leading to high pest populations. This is out of the farmer's control.

But what other factors can cause insect mortality? We observed during long dry spells that larvae became flaccid. John Hilditch found that scarab larvae could lose about half their body water, and they would still recover if placed in moist soil. But if they lost more than this, they could not recover. The soil air is saturated except below the wilting point, which for each soil is the moisture content at which plants can no longer absorb water from the soil. As the moisture content falls below wilting point, the soil atmosphere becomes increasingly dry. Scarab larvae have no control over water loss through the cuticle in these conditions. In a few days, sufficient water may be lost to kill larvae. Large numbers of larvae are lost in this way, and in dry weather eggs may fail to develop because the soil is too dry for the absorption of water essential to their hatching.

How often do tableland soils fall below wilting point? John Begg calculated that in New England, over a four-year period, soils at a depth of 7.5 mm were below wilting point for an average of over 20 days in December and January. Here then is another common cause of mortality in soil insects — desiccation of larvae which cannot burrow deep into the soil or construct a cell to tide them over the dry periods. Phalaris/clover pasture has a similar pattern of drying out to a depth of about a metre, but native pasture does not dry out to the same depth.

At the other extreme, prolonged rain will cause heavy insect mortality in any soil which remains sodden for a week or more. In the autumn of 1978, this

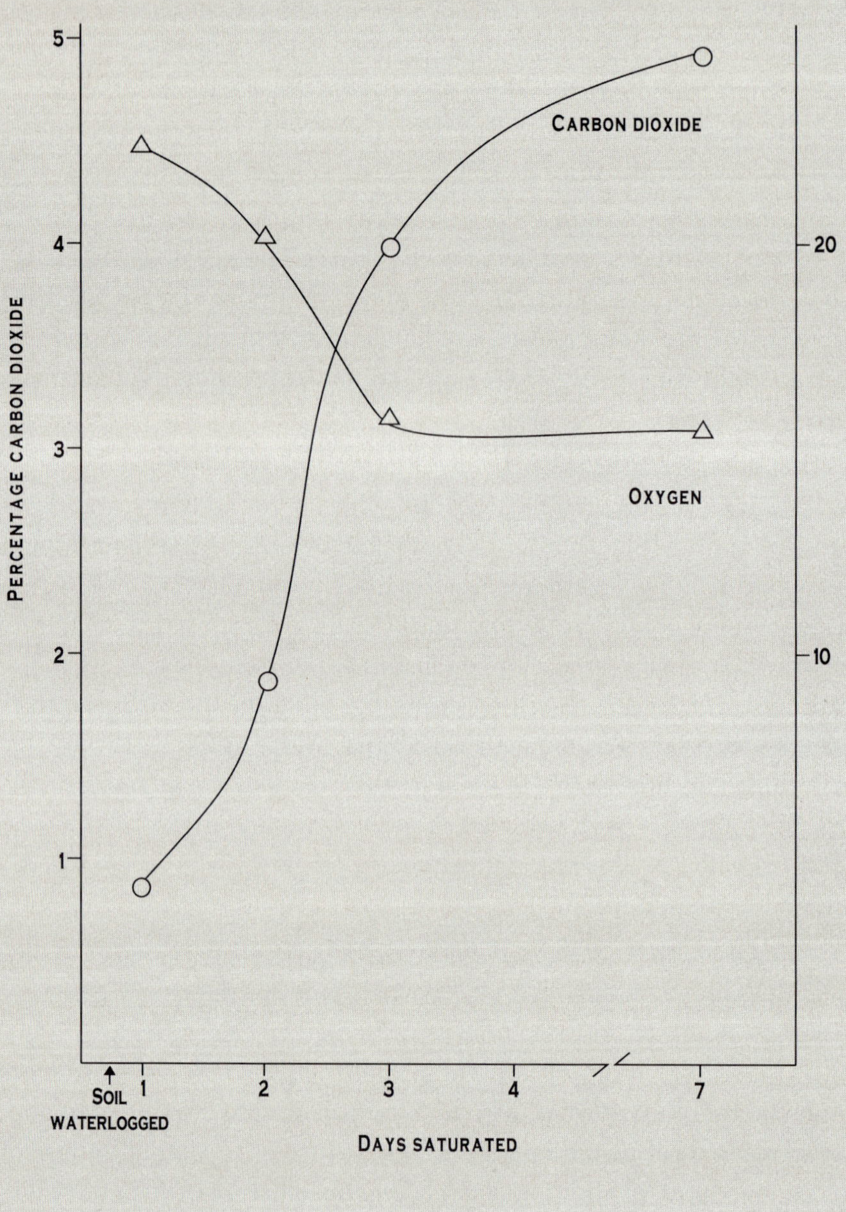

Figure 14

The percentage of oxygen (△) and carbon dioxide (○) in soil air measured over seven days following saturation of the soil under grass.

condition prevailed long enough for vast numbers of pasture scarab larvae to come to the surface in the water. Oxygen content of soil air, and oxygen dissolved in water, are deficient and carbon dioxide concentration becomes too high for insects (see fig. 14). Magpies, ibises, herons and many other insectivorous birds feed day in and day out on those insects that come to the surface in the prolonged rains.

Wide fluctuations in temperature and moisture affect all animal populations. Warm-blooded animals survive the unfavorable periods better than the insects, which succumb in great numbers. As we saw in table 1, most insects lay many more eggs than could possibly survive. Unfavourable weather conditions keep the numbers of most insects down, but weather is quite variable from year to year and the consequence is a fluctuation in the numbers of most insect species. These ups and downs were discussed in the previous chapter, and we concluded that in pest control we should always take a long view of fluctuations in pest populations. Farm management should aim to enhance the natural controls where possible.

While we cannot change the weather, the crops grown and the management of pastures may alter considerably the microclimate in which insects live. Vegetation cover modifies the temperature of soils, and the increased plant cover from improved pastures reduces soil temperature compared with sparse native pasture (see fig. 15). On the other hand, the heavy grazing of palatable grasses and clover may denude pastures, thus increasing soil temperatures by many degrees.

Trampling by stock has an effect on soil porosity, and in wet weather trampling may compact soil. The powdery soil from trampling by stock decreases the percolation of rain, and runoff and erosion may be greatly increased by heavy stocking.

Good farm management requires careful consideration of the effects of farming practices on potential insect pests. Pasture improvement usually increases the quality of food for insects as well as for domestic stock. We should not be surprised when indigenous species increase their numbers when sown pastures provide an abundance of food and also a more congenial microclimate.

PHOTOGRAPH 4

During the first 100 years of pastoral production, burning off unpalatable grassland in late winter removed the litter under trees, exposing soil to excessive solar radiation. High soil temperature and low soil moisture were often lethal in summer to the eggs and young grubs of scarabs.

PHOTOGRAPH 5

Trampling by cattle and sheep causes compaction of moist soil and pulverising of dry soil, resulting in sheet and gully erosion where the grass cover is insufficient to hold the soil together.

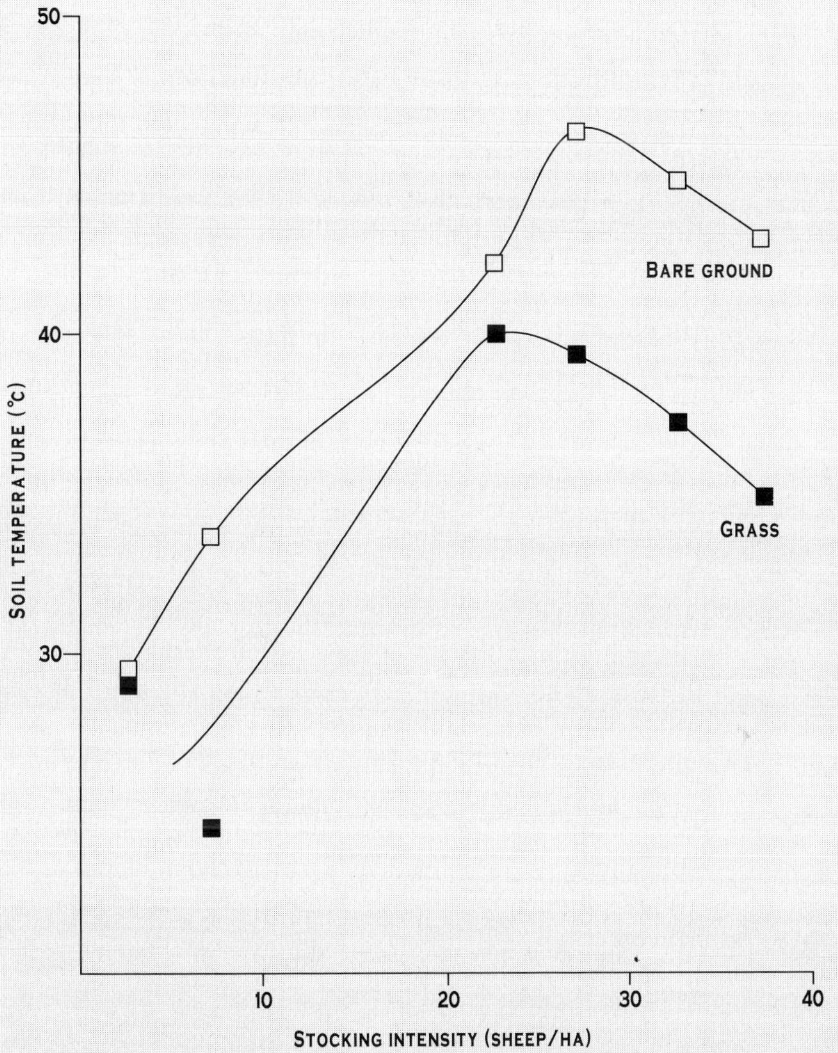

FIGURE 15

Soil temperatures at a depth of 2.5 cm measured on a hot day in pastures grazed by sheep at a wide range of stocking intensities. Temperatures under grass (■) and under bare patches (□) are affected by pasture cover and trampled soil.

Microclimate is one of the factors regulating populations which the farmer and grazier can modify to reduce insect pests. We suggest that the detective work done on scarabs, identifying the exact causes of environmental stress, be extended to other potential pests. With such information a landholder may yet live in peace with insects kept in their place. In the previous chapter Graeme Baker was quoted on wingless grasshoppers. Heavy grazing of shallow hilltop soils provides bare soil in which grasshoppers lay their eggs, especially on the hot northern aspects. Landholders could reduce outbreaks by controlling the grazing of these sites and planting them to bushland to shade the soil.

If a list of pasture pests on the tablelands was compiled, it might be frightening. If the potential reproduction rate was added for each species, it would immediately become apparent that already there are effective natural controls. After all, we know that the species on that list are pests only infrequently, but why are the species all kept so far below their potential numbers?

Part of the answer has already been given in this chapter — environmental stress kills a great proportion of the immature stages of these insects before they do any harm. The second part of the answer is that the remainder of the population is being killed by a vast array of predators, parasites and diseases. Let us introduce these species of birds, insects, spiders, lizards and insect pathogens before you make a judgment on pest control.

5

Predation: an Insatiable Appetite

Like a miniature seal the green maggot wriggles ponderously along the stem. Gripping with its broadened tail end it suddenly rears up and, with stabbing movements of the front part of its body, investigates the immediate surrounds. Finding nothing, the maggot moves further up the stem of the plant and repeats the arc of head thrusts. Here, near the terminal bud, are clusters of minute aphids sucking juices from the plant and suddenly the maggot has one. It holds the hapless aphid aloft in its mouth, like a seal balancing a ball, and slowly ingests the prey.

Close by a hover fly darts from flower to flower, hovering motionless before shooting away like a silent helicopter. Sometimes it lands and sips nectar from a flower, refuelling before a remarkable vertical take-off. Occasionally, while hovering, the fly rubs its legs together, perhaps to remove unwanted pollen grains inadvertently collected from the flowers. Then the undercarriage is neatly tucked away, aerodynamically safe and sound, and the fly moves off in search of flowers or perhaps an aphid-infested bud on which to lay an egg.

Remarkably, the corpulent maggot munching lazily on aphids amongst the stems, leaves and buds, and the aerobatic hover fly in its gold and black colours are one and the same. After a life of gross gluttony the maggot undergoes a profound metamorphosis during the pupal stage to finally emerge as a hover fly. The highly mobile hover fly or syrphid with its sophisticated eyesight seeks out sources of nectar, or sugary honeydew excreted by aphids, on which to feed. After mating the female lays the eggs. From these the maggot larvae hatch to once again menace the plant-sucking aphids.

Growth is the prime function of the larva and hence its great importance as a voracious predator of aphid pests. Dispersal and reproduction are the duties of the adult fly and hover flies breed quickly in spring, their numbers remaining high as long as aphids are still present. Syrphid numbers decline in autumn as the aphids become scarce with the onset of cold weather and shorter days.

Syrphids are not the only predators of aphids. Fortunately for farmers and gardeners, several species of ladybird beetles which we see on roses, broccoli, fruit trees and lucerne — in fact wherever aphids are present — are also aphid hunters. Both larval and adult ladybirds, known technically as coccinellids, actively seek out aphid prey. The legs of the larvae are quite strong, but in the adult are less conspicuous, giving these bright little beetles the appearance of a toy battery-operated tortoise running smoothly along. The gay red or orange colours, with black spots or blotches, belie the predatory nature of ladybirds, but the larvae look the part, lurking ominously amongst the vegetation in their dark spiny armour.

A species of ladybird known as *Rodolia cardinalis,* introduced from Australia, was used most successfully as an agent of biological control of the pestiferous cottony-cushion scale in the citrus orchards of California. Do we appreciate our good luck in having such a great variety of benevolent predators of insect pests here in Australia?

Of course, their predatory activities have no function other than to ensure survival of the species, but mankind gains such incalculable economic benefit that it is tempting to imagine that these enchanting creatures are under firm legal contract to minimise pest problems in our farms and gardens.

Life is not always easy even for Australian predators. Imagine living in a world in which your food supply fluctuates from surfeit to famine. Unless we understand that predators of insect pests need a continuous supply of food we will never learn how to encourage natural controls of pests. Low population levels of pests must be tolerated to provide the food on which predators can survive and increase their numbers. It is futile to aim for pest extermination. Insecticides should never be used while pest populations are low and damage is negligible. When you spray your roses at the first sight of aphids, you make it impossible for the predators to build up their populations. Next time, first look for predators and you may witness the drama of natural control in your own garden.

With luck you may find the nest of a colony of paper wasps, family Vespidae, suspended from a branch in a bush. The platform nest is started by a fertilised female who survives the cold winter, lays the first eggs, and rears the helpless larvae until they mature to become winged workers. The workers then assume the responsibilities of building new cells, in which the queen lays eggs, and of foraging for insects to feed the larvae. The worker wasps masticate the insect food into a pulpy mass before it is served to the larval wasps. Defence of the hive-like colony is also the duty of the workers and their sting is reported to be painful if the nest is accidentally bumped. But generally inquisitive humans are intimidated enough by the waving antennae of a defensive line of wasps not to test their last reserves. In autumn, male and female wasps are produced and mate. The males, sterile workers and the former queen die with the onset of winter, but the females hibernate and emerge the following spring to initiate new colonies.

Everyone is aware of the seemingly hypnotic attraction of some beetles and moths to an electric light on a warm summer night. Less conspicuous than the frenetic, fluttering moths are the green lacewings which sit quietly, bemused, in the radiance of the light. Pastel green with lace-veined wings and glowing eyes, they must be one of the most attractive carnivores alive. Their larvae feed on lerps, scale insects and the like. One larva can consume up to 400 aphids before pupating. The ingenious larvae have hairs on their upper abdomen with special hooks to which they attach small fragments of bark or other debris to effect an almost perfect camouflage.

Other members of the order (Neuroptera) have a different colouration and shorter antennae or even a completely different body form to lacewings. The mantispids look like praying mantids, but retain the typical superb lace-like wings.

Larvae of one family, the antlions, bury themselves in dry soil and dig a small pitfall trap. The antlion lies in wait below the pit to seize any ants or other insects which stumble into the trap. So adapted are they to this life in the soil that their body hair faces forwards to facilitate backward movement through the soil and presumably to assist in the struggle with prey by providing a grip in the loose sand. Antlions have the strangest of personalities. If dug from their sandy abodes for examination, they are extremely docile and disinterested. But, just as you decide that the captured specimen must be ailing, it will make a backward lunge so quick that it is almost missed. The twitch is then followed by a long anticlimactic period of doleful immobility.

Like a stationary front-end loader, in miniature, the antlion lies leadenly on the sand, its huge jaws resting before it. You can only take pity and bury it, so that it may one day emerge as an adult antlion fly.

Meanwhile, hour after hour, day after day, the antlion keeps a lonely vigil, waiting for the frantic vibrations which indicate an insect in the pit and, with skill and luck, the repast of the lion. Their enterprise and patience are truly admirable.

It is during the warm season that the call of the Australian bush beckons irresistibly and we are drawn to shady picnic areas. Invariably these weekend outings take us to a body of water where the breeze is humid, cool and relaxing. And it is near these streams, lakes and ponds that damselflies delicately flutter or perch quietly on waterside reeds and grasses. Larger dragonflies hover restlessly above their reflections, resplendent in red, blue, yellow or green, often striped with black.

While watching these insects, half an hour can pass like a few minutes as a particularly robust dragonfly puts on an amazing air show, with turns of speed and rapid changes of direction which leave you awestruck. The huge compound eyes of the dragonfly befit such accomplished aviators, giving somehow the impression of a begoggled pilot. On its landing the illusion is not shattered. The four unfolded wings of the insect glisten in the sun as might the metallic rotor blades of an earth-bound aircraft.

While the damselfies and dragonflies hunt for mosquitoes, horseflies and other insects on the wing, less spectacular dramas are occurring below the water-line. Here, where the dense, murky water filters the light and deadens sound, a long-legged, articulated creature crawls along the silty bottom in search of crustacea to grasp and consume. After a year, or maybe several years, of this aquatic existence the nymph hauls itself from the water and moults to become a mature dragonfly complete with net-veined wings — an instant replay of millions of years of evolution as primordial forms emerged from ancient waters to evolve, eventually, into completely terrestrial organisms. Dragonflies and damselflies effect this transformation within a single life cycle.

Robber flies (asilids) are also familiar predators in the Australian environment. These are bold and strong flies with a stout thorax or middle section, but an elongated abdomen. They have muscular legs for grasping prey and, like all self-respecting aviators, they often sport a thick moustache. While adult robber flies capture winged insects, the larvae dwell in the soil or decomposing wood and attack soil-dwelling organisms. These asilid larvae are an important natural enemy of scarab larvae under pasture.

It is easy to overlook the predators that burrow, shovel and squeeze their way through the sand, soil and mud of the earth. The soil environment harbours a host of minute animals and micro-organisms that utilise plant and animal remains and break them down to form simple nutrients that are absorbed by plant roots and hence recycled. Herbivorous forms, such as the scarab larvae of the Christmas beetle and related species, graze the roots and ingest other organic matter in soil. And where there are animals in the soil there will be predators hunting them down.

Adult and larval ground beetles or carabs, with heavy mandibles, are conspicuous predators of animals in soil, dung and plant litter. Some wire-worms, which are larvae of click-beetles, are also predators in soil and litter. Both carabs and elaterids are elongated, usually black, beetles. The carabs often secrete an obnoxious substance if handled, while the click-beetles are so named because there is an articulation in the beetle's body which gives an audible click and propels the beetle some distance into the air. Many rove beetles (staphylinids), recognisable by the short truncate forewings, are also predacious both as larvae and adults.

What a variety of organisms are to be seen in a whole range of habitats, spanning the claustrophobic crevices of the soil to the freedom of the air and the cool waters of freshwater streams. Yet the fat maggots and daredevil dragonflies, cryptic antlions and colourful ladybirds, the praying mantids and mantispids, not to mention the professional assassin bugs and the carabids of the soil, share one common denominator. Each and every one is a predator.

Some feed on mosquitoes, others on aphids, scale insects, scarab larvae, grasshoppers, crickets, horseflies, caterpillars and moths. The list is seemingly endless. The instinct to kill and consume is so strong, the appetite so insatiable, that some predators do not even distinguish between 'them' and 'us' and cannibalise their own species if the chance arises. But the predators, despite their gory way of life, elicit great interest. The more we learn of these creatures, the greater our gratitude, for predators have helped to keep a check on the numbers of other insects for millions of years. Their murderous ways save farmers and graziers thousands of dollars and home gardeners much distress and frustration.

If you think you have a pest problem, imagine its magnitude without the benevolent predators which daily undertake pest control without any effort on your part. Before reaching for insecticides, the short-term panacea, think of the harm they do to the ladybirds, lace-wings and others of their ilk, which inevitably destroy large numbers of pests year after year. At the very least, use only those insecticides which are fairly innocuous to predators.

Think not only of the few predacious insects described in this chapter, but realise also that there are many other predators wherever insect pests occur. There are so many that some have yet to be named and described by taxonomists. It is little wonder that so few insects survive to reproductive age. A newly-laid insect egg has odds of perhaps one in 100, maybe one in 20, but very rarely an even chance of reaching parenthood. It is never a certainty. Most insects lay many eggs so that a few will reach maturity and reproduce to perpetuate the species. That is to say, because there is usually no parental care and life is such a gamble for the young insect, natural selection has generally favoured the more fecund females, able to lay many eggs or, if you like, able to take more tickets in the lottery of life. And the losing tickets belong to those insects that suffer from extremes of weather, or succumb to disease or parasites, or that go down the throats of insatiable predators.

If farmers and gardeners care to look for the useful insect predators described here they will undoubtedly find them. Most of them, or their close relatives, are found throughout Australia. Indeed, similar species occur throughout the world. However, if farmers and gardeners wish to encourage them as a means of successful biological control they will need to consider carefully the nature of the delicate relationship between predator and prey, as well as the essential habitat requirements of these useful and delightful animals. Insect predators need food, so leave a few pests and innocuous insects for them to feed on throughout the year. Provide flowering trees, shrubs and herbs in bushland and on farms, and leave plant litter on the ground for shelter.

Of course, all predators are not insects. In the following chapters we shall talk about birds, lizards and spiders and their habitat needs.

6

Parasitism: A Life of Luxury

Give a pest a name that is easy to pronounce, then tell people how to recognise it, and there will be reported sightings from all over the district. While this is understandable, it is unfortunately equally true that people overlook the insects that they do not recognise. Most people are oblivious of the great number of parasites and predators that are quietly going about their useful activities in woodland, in pastures and in gardens. Let us introduce you to some of these species, so that your farming and gardening may be carried on with an awareness of these natural controls.

In summer you might find a brightly-marked black and yellow wasp caught in the radiator of your car. The glassy wings and constricted waist suggest correctly that it is a wasp, but it is large-bodied and over 2 cm long. In fact it is perfectly built for its peculiar role in family life. It is a male thynnid or flower wasp. Perhaps you have not met the female of the species *Hemithynnus hyalinatus*? She is only a little smaller, but wingless, and she spends most of her time burrowing in the ground in search of scarab grubs. She lays one egg at a time on a fat grub, and after doing this once or twice she comes to the surface and waits for a male to be attracted by the odour she exudes at this time. If she is unsuccessful in attracting a male while in the soil with only her head above the surface, the female will crawl up a grass stalk and adopt the mating position there.

The females of this species require fertilisation after only one or two eggs have been laid. She is fed by the male while copulating. The behaviour is bizarre, but quite essential. Females cannot fly to sources of nectar, so the male regurgitates nectar from his stomach to feed the female during copulation. Pairs remain coupled for about half an hour, and during this time the female strokes the male's abdomen to stimulate the flow of nectar. Mating always takes place on the ground, but sometimes there is a post-nuptial flight. However, the male's orientation is so perfect that after a flight of 100 m he returns to the same place and safely deposits the female.

She immediately burrows into the soil in a shady spot and searches for another scarab grub. A large grub or pupa is chosen, and stung to paralyse it; it is then left in the ground after a single egg has been laid on the host. What a life of luxury for the wasp larva! For about a fortnight it feeds on the preserved grub; there is enough food for the wasp larva to grow to its full size. Then it spins a cocoon in the soil, and here it remains as an inactive pupa for ten months. An adult thynnid wasp emerges from the cocoon in spring.

James Ridsdill Smith, who studied this wasp at Tilbuster, near Armidale, has found that this species parasitises grubs of five large scarab species. The sobering thought from this study was the fact that male wasps needed to feed on nectar every day, and they would fly only 200 m from nectar-bearing trees and shrubs such as eucalypts, blackthorn, sticky daisy bushes and tea-trees. This most useful wasp was not found at greater distances from trees. What a great pity that there was not an awareness of this fact when land was cleared for pasture improvement.

Most parasitic wasps, male and female, feed on nectar from flowers or honeydew from scale insects. The wasp larvae, on the other hand, mostly feed on other insects or spiders. Wasps are the most important group of insect parasites helping to control pests which damage plants. Apart from bees and ants, which belong to the same order, there are many thousands of species of wasps in Australia. In most species the wasp larva lives a life of luxury. The egg may be laid on the host, sufficient to feed the larva until it pupates. Or the wasps may feed the larvae in cells of various shapes in all sorts of places. However, you cannot count on these beneficial wasps unless there are flowering plants nearby on which the adult wasps can find nectar.

Next time you see on the roadside a blackthorn, heavily laden with pale cream flowers in the late summer, stop for a moment and count the wasps feeding on the sweet-scented blossom. You could easily find ten scoliid or hairy flower wasps with metallic blue wings and body and bright yellow head, as well as a few individuals of other wasp species. Walk out into the pasture near such flowering shrubs, and you will see dozens of the metallic blue *Scolia* wasps flying or crawling over the pasture. They parasitise scarab grubs in the same way as the thynnid wasp already described. Walk some distance away from the source of nectar and the wasps will not be seen.

Many species of thynnid and scoliid wasps occur in Australia. There can be little doubt that their numbers have declined as the sources of nectar have been destroyed. Yet there is an abundance of food for the wasp larvae — scarab grubs damaging the pastures. Where is the sense of talking about introducing biological controls from overseas when the indigenous controls still persisting here are restricted by the thoughtless development of inhospitable pastures? Possibly scientists are rushing ahead so fast that they have forgotten why they are doing research.

Have you ever watched a solitary wasp building a mud cell on a lazy summer's day? Every few minutes she flies through the window to a space between two bricks on the wall, or a dark corner of a bookshelf. The short bursts of high-pitched humming as she moulds the wet soil into shape will lead you to her building site. When the cell is formed she will return again and again, each time carrying a spider to pack into the cell. An egg is laid in the cell, on the paralysed spiders, and the cell is carefully sealed like a mummy's tomb. Other solitary wasps drag caterpillars to their nests. If you have seen this activity, you will realise how much energy goes into providing that life of luxury for the wasp larva. And remember that all that energy comes from the nectar she has to search for on a flowering shrub.

These sphecoid wasps, which fill a mud cell with prey and then lay an egg in it, are important in regulating the populations of a wide variety of insects, including caterpillars, grasshoppers, bugs, flies, beetles and sawflies. Many construct their cells in the ground. When collecting prey, they are industrious and knowledgeable. Usually they collect only one species. When we watched the slender red-brown wasp with the yellow-banded legs and body she returned again and again at about three-minute intervals to store spiders in a cell, each time with the same species of spider. When one cell was closed, she began building another against the first cell. One wasp will destroy many pests in a short time and you should be glad if these wasps are common in your garden. In a puddle of water we watched three wasps collecting mud at the same time. One species scraped the soft mud and flew off, but the other collected its mud and skated several times across the water before flying off, satisfied that the mud was wet enough.

There are innumerable small and tiny wasps which parasitise insects of many species, each wasp usually attacking only one species. Many pests like aphids, scale insects, lerps, moths, flies and beetles are attacked in the egg, larval, pupal or adult stage. It would be a pity if our agricultural practices did not encourage these parasites of insect pests. Virtually all adult wasps require nectar or honeydew — so providing native flowering shrubs is one way to do this.

Jeanette Conacher tells how some of these wasps have been introduced to Western Australia to control insect pests such as army worm and scale. Anthony Healy and Courtenay Smithers describe and illustrate, in their book, several of these parasitic wasps, and the usefulness of this group of insects is shown, too, in F. Hockings's book *Friends and foes of Australian gardens.*

Ichneumonid wasps (of which there are 1244 species in Australia) are slender wasps which mostly lay their potentially deadly eggs in caterpillars. The larval wasp feeds on the tissues of the unfortunate caterpillar, and when fully grown leaves the host's body to spin a cocoon in which to pupate, usually on the back of the caterpillar. Braconid wasps (800 Australian species) parasitise the larvae of moths, butterflies and beetles as well as aphids and scale insects. Chalcids (3646 species recorded in Australia) are small to minute wasps, some of which parasitise insects, others causing galls on plants.

In chapter 3 the value of scelionid wasps in complementing pesticide spraying to control locusts and grasshoppers is outlined (see Graeme Baker 1983). One *Scelio* species typically causes 1–4, 10–30 and 70–90 per cent egg mortality in successive generations of the Australian plague locust within a season. Some 445 species of scelionid wasps occur in Australia.

Related to these hymenoptera are some species which feed on plants, and a few are serious pests. The sawfly is one of the destructive insects found on eucalypts. The damage done by these unpleasant grubs is described in chapter 12. Another sawfly pest is the introduced pear or cherry slug, which also skeletonises the leaves of ornamental trees like flowering cherries.

Many flies lay their eggs on or near insect pests, and the fly larvae enter the body of the pest and destroy it over a period of weeks. During this time the pest may remain active. Bee flies are large flies which may be seen hovering and emitting a low hum. The female flies do not feed, but will lay up to 10 000 eggs in a knothole in a dead tree or wooden fence post. The larvae which hatch out are resistant to desiccation and hairy, so that they can float in the wind. These minute bee fly larvae enter between the segments of wingless grasshoppers and plague locusts, where they develop by consuming the ovaries and fat bodies of the host. When fully grown the larva chews its way out of the dead grasshopper. In Canada old wooden fence posts are placed in pastures where no dead trees remain, so that bee flies may lay their eggs in the holes in the posts. Throughout Australia many outbreaks of locusts and wingless grasshoppers are controlled in early summer by bee flies, tachinid flies, wasps and parasitic nematode worms.

Insect populations are greatly influenced by pathogens, which under suitable conditions will cause heavy mortalities. Richard Milner has found that corbies or underground grassgrubs (species of *Oncopera*) can be almost wiped out by disease when dense aggregations occur in wet weather. Two diseases, one a protozoan (single-celled animal) and the other a virus, are thought to be responsible. A range of pathogens is recorded for scarabs in New England, mostly fatal to the grubs which are infected. Eleven pathogens are known to occur in one scarab species which occurs between Ebor and Dorrigo. Dense aggregations of some scarab grubs have been observed to collapse dramatically when milky disease (a bacterial infection of the blood) occurs in favourable weather conditions. Fungal diseases, which fill the larva with their filaments and mummify the cadaver, also occur.

One of the most striking examples of disease causing a collapse of insect populations is the effect of the fungus *Entomophthora* on grasshoppers. During one recent summer a pasture which had been closely grazed was plagued by an abundance of a grasshopper with the noble name of *Praxibulus*. The short grass was alive with them, and then suddenly the numbers dropped although the weather was perfect for grasshoppers — hot and dry. Then we noticed the dead *Praxibulus*, all of which had crawled up the stems of the stalky *Stipa* or bamboo grass nearby. Each grasshopper died while clasping on to a dry stalk, a weird response of hoppers infected with this fungus. Lucky for everyone, because the spores of the disease are widely dispersed from the elevated corpses. Who knows how serious a pest this grasshopper would become, were it not for the biological control exerted by the associated fungus?

Many attempts have been made to propagate disease pathogens to spread them in outbreaks of insect pests. One of these pathogens is readily available from retail outlets in Australia. It is a culture of a bacterium which can be spread around gardens, and it will effectively control cabbage white butterfly caterpillars. Unfortunately the research and development of such controls is expensive. It remains only to remark that many such diseases occur naturally in Australian pastures and gardens.

If you are to take advantage of these parasites and diseases, you should understand clearly that the beneficial organisms — parasites and pathogens — can persist on your property only if a host is present. If you use a chemical insecticide each time you see an aphid or a grasshopper, you will never see the miracle of a disease or parasite overwhelming a pest in your garden or pasture. Disease outbreaks develop mainly in favourable weather when pest density increases. Parasitic populations cannot build up from nothing. Low densities of pest species, which cause little damage, should be left alone to act as a reservoir of hosts for natural controls, such as diseases and parasites, that can spring into action when pest numbers increase.

7

How Many Birds in Your Garden?

If you have flown over a city or town, you will have seen the streets and houses crowd out the green grass and the colourful trees. Fly lower and you will discern how the severe structure of tile and brick, of concrete and asphalt, is softened by the individual efforts of home gardeners. You will see the cool shade of trees falling across watered lawns. The dense foliage of trees shimmering in the cool breeze would seem to offer shelter if you were a bird.

This is precisely how all the birds that live in town find a home and a feeding ground. They are not attracted by the streets or houses, nor are they frightened by these structures of ours. But they seek the shelter of trees in the first approach, then feed amongst the foliage and flowers, or swoop down to the lawn or the vegetable patch.

While you sit quietly on the patio a tiny flock of minute yellow-rumped thornbills will fly fearlessly towards the house, onto the fence rail, and then to the mown grass. As they hop daintily around they are picking insects from the lawn. Their sweet animated song is said to be a contact call to keep the flock together, as the birds tend to move towards the caller. They utter a sharp call in flight. When they fly all together into a eucalypt tree or your rose garden, they will feed for a while on aphids or any soft-bodied insects. Usually they will stay while there is food enough.

Those keen gardeners who boast that there is not a pest in their garden might count the birds in their tidy garden and compare that with the number of birds usually found in the rampant garden of their lazier neighbour. In a small suburban garden half a dozen thornbills or other insect-eating birds will keep some pests under control. This does not mean you will never see an aphid on your roses. Quite to the contrary, it means that there are enough insects in your garden to attract the birds. While they are feeding in your garden daily, your hope is that they will keep pest numbers so low that damage is negligible. Your choice is between a natural control and the use of an insecticide which will kill all the insects on your rose bushes at that time. When the thornbills come looking for their daily feed, and find nothing to eat, they will move on to some other garden. It may be many months before they return to your roses, and you are left without their graceful beauty and without their beneficial feeding. There are several other killers of aphids, for example tiny parasitic wasps and predacious ladybirds. These will disappear too, either because the insecticide spray was toxic to them or simply because there are no aphid prey.

Wattlebirds announce their presence with a raucous determination, and then you see them in the outer foliage of gums or ornamentals. Early in the morning they feed lazily, flying from treetop to treetop, searching for the insects that make up part of their diet. Unlike the thornbills, which feed mainly on the ground, these red wattlebirds are birds of the outer foliage, where they search out bugs and beetles. Feeding in pairs or small groups, they follow one another in an animated flight around the garden. They are common in suburban gardens, and can be seen at any time of day. When feeding young in the spring they are fearless and oblivious of your approach, squawking loudly as if to say 'don't you dare!'. At this time they feed ravenously on nectar in *Banksia*, *Grevillea* and many other ornamental flowering shrubs. At times they feed also on soft fruits in orchards, but the harm they do is far outweighed by the good. They feed their young mostly on insects. The mature bird has handsome red wattles nearly an inch long. The male's varied calls are all harsh, and it clicks its bill aggressively at its fellows. Females make a more musical 'phew, phew'.

Although an inconsistent visitor, the noisy miner or soldier bird will visit gardens in great flocks in some years in response to feeding trays with bread and honey or sugar. They will become quite tame and never seem to squabble amongst themselves, no matter how crowded they may be on a feeding table. They are described as garrulous, with two calls — a repeated single note or a loud, rapidly repeated 'wick-wick-wick'. Insects are prominent in their diet, including lerps, beetles and caterpillars procured among the leaves and blossoms and on the ground. They also feed on nectar from eucalypts in flower.

There is a bird you'll hear again and again before you see it, because its colour is quite dull grey. But one day your eye will catch a coquettish movement — a little bird twisting on its legs and at the end of each arc spreading its erect tail in a delicate fan. With the light shining through the separate feathers it is an exquisite sight. There is delight for any gardener when a solitary grey fantail performs for your special pleasure. How else can you interpret the spirited performance of a bird that is fearlessly singing and dancing when you are the only audience? Practising? Surely not. Expressing a sheer joy of living? You would be possessed of a dull soul indeed to think it is instinct alone. It sings a sweet, twittering song while performing tirelessly and seeking flying insects, which are its only food. In the breeding season this little bird has a more beautiful song — high-pitched, upward soaring and very melodious, reminiscent of a violin.

A larger, more handsome cousin in many Australian gardens is the willy wagtail. This jet black bird with white chest and eyebrows prances around in similar fashion on the ground, feeding on insects of all kinds. It seems to seek the company of cows grazing or lazing in the paddocks, strutting along their backs, dropping to the ground and alighting again on the backs of the animals. The first time this delightful bird hops along a fence-rail towards you, peering up at you and whipping its tail from side to side, crying 'sweet pretty creature' you will know the meaning of the well-worn phrase 'communing with nature'. You could not at that moment help but feel guilty of depriving it of the insects it feeds on. Its alarm call is described as a chatter resembling the rattle of a half-empty box of matches. It is exceptionally tame and will nest in sheds, under verandas and in vines growing about houses, or on the horizontal branch of a fruit tree.

Another very confiding bird which will visit houses, seeking spiders in windowframes, is often mistaken for the willy wagtail. The restless flycatcher is black with a white breast and throat, but without white eyebrows. The feathers of the crown can be erected into a small crest. Flies, moths and other insects are taken on the ground. While it searches it can hover for some time at less than a metre above the ground, uttering the grinding sound which has led to its other common name of scissors-grinder. Its other call is a long, loud, repeated single note 'pee-pee-pee'. The nest is built on a fork in a dead branch of a tall tree, bound with cobweb and lined with fur. Although able to adapt to clearing of bushland, it is not as common as the willy wagtail.

Have you ever seen a dozen silvereyes in a coppice of silver birch trees? In autumn, with the leaves partly fallen and the sparse foliage a pale yellow, the slender twigs of these white-stemmed trees scarcely bend as the grey-breasted silvereyes flit in small flocks among the yellowing leaves. Each bird is no larger than a single leaf of many of the trees in which they forage. As they flit among the outer twigs they call to one another in a song described as fragile but melodious and sustained. They feed on all manner of insects during summer, including leaf beetles and grubs. In spring some bring their young to feed in a flowering shrub growing against the kitchen window. They are grey-green and when the *Feijoa* is in flower they feed on the fleshy petals of this shrub. In winter, migrants from Tasmania with dull colours and brown flanks swell their numbers.

Currawongs in a flock act like a busload of tourists, following one another from tree to tree. They love the tallest eucalypts. When they fly down to feed on ripening apricots or figs, they return frequently to the tall trees. All the while they sing together in chorus, with a wide repertoire of full-throated melodious songs — 'curra-wa, curra-wong' and a long drawn-out 'chilla-wong'. From high up in the treetops the performers do not need the acoustics of a concert hall. Orchardists and poultry farmers do not love them, but in gardens they feed mainly on large insects and some fruit. These pied currawongs visit towns in the cool weather and return to mountainous forests to breed in spring.

The bad side to currawongs is their habit of destroying the eggs and nestlings of other birds. This is distressing for nature lovers. An increase in currawongs where firethorn, hawthorn, privet and camphor laurel bushes are planted may depress the numbers of other more useful birds.

The changing seasons bring a change of pace in the daily activities of all the birds. People who work a regular day throughout the year can sense this change of pace in the birds in their gardens. In September every year there is feverish activity among all the resident birds. Courtship is a prelude to the ceaseless activity of building a nest and rearing young. At this time more than any other you will appreciate the special habitat that each species needs. When you are planning a garden, bear in mind that the trees and shrubs will be examined by the birds as they search for a safe place to nest each spring.

Eucalypts are a favoured tree for many species to nest in, but a clump of tall eucalypts will be preferred over an isolated tree. You wonder how many sites are examined by the cock and hen before they settle on the one they will use. A pair of wattlebirds will choose a fork in the thicker branches of a gum tree, perhaps a fork where three or four branches arise together. It will be at the end of a vertical smooth stem that cats cannot readily climb, and away from horizontal limbs that overlook the nest. Once the nest building commences the birds are quite secretive about their comings and goings.

When a nest is full of demanding chicks, feeding these is the sole preoccupation of the parent wattlebirds, too busy now for the chasing of one another through the shrubs in flower. Cock and hen feed quietly at this time, snatching food and flitting back to the nest. However, they make a raucous noise away from the nest. If a pair sees an intruder, the bird on the nest may fly quietly to a tree some distance away and set up a great noisy chattering to attract attention. Its mate, instead of flying in to the nest as before, will join in the noisy diversion and both will keep away from the tree in which their nest is situated. The young birds will not give away their position.

In the same gum tree there may be the very untidy nest of a pair of yellow-rumped thornbills. The structure is large for so small a bird, with an inconspicuous entrance low down. The size and untidiness arise from the cup-shaped platform above the nest. No explanation (except perhaps a false nest to confuse predators) can be given for this unusual feature of thornbill nests. The nest is built within a tangle of thin dead twigs on the outer branches of a tree or a leafy dangling branch. Nests are built by the males, who keep the secret of the superimposed nest to themselves. Males do not feed the hen who is sitting on eggs, but both cock and hen feed the young; even the other thornbills in the flock will assist in the feeding. Their comings and goings are quite uninhibited, and this is one of the friendliest of suburban birds.

Another friendly bird that nests unafraid in suburban gardens is the superb fairy-wren. The ballet-like gestures of the brilliant blue cock and the gentle grey-brown hen draw attention to the chosen nesting site, a low bush or creeper near the ground. Seldom more than a few feet from the ground, these nests are easy prey to cats, lizards and other predators. Time and again delight in observing nests of wrens with eggs and chicks has changed to anger on finding the nest broken on the ground. This occurs not only in towns, but in the bush as well. Feral cats, that is domestic cats gone wild, are a threat to wildlife of many kinds. A tangle of dense, thorny scrub is the only safe place for birds such as wrens, double-barred finches and other small species which choose to nest close to the ground. We discuss the disappearance of suitable undergrowth on farming and grazing land later. However, in your own garden you can attract these birds by leaving an untidy hawthorn bush or tangle of honeysuckle on a back fence, in which wrens will nest. These wrens will feed on insects in orchard and vegetable garden.

The common Australian magpie has adapted perfectly to urban development in towns and cities. It nests in tall eucalypt trees, the nest resting precariously in the outer foliage, built on slender horizontal branches. The female builds the nest of twigs and grass, and lines it with wool and fur. Three or four eggs are laid once each year, but chick mortality is high, and usually only one chick survives to adulthood, even in a sheltered garden. Two or three nestlings blown out of a nest are often found on the ground, too young to survive.

In chapter 8 the territorial habits of magpies are described. Your garden may be part of the permanent territory of a breeding flock, and you will observe that they defend this territory all year around. Their melodious carolling begins before daylight and may sometimes be heard throughout the day until after dark. They harmonise as a group, with a fine repertoire of songs, but they also sing solo as they check the boundaries of their territory. Can you discern these boundaries in your own and your neighbours' gardens? Other large birds which threaten young magpies will be chased out, but only when young are being fed on the ground will mudlarks, starlings and a few other birds be chased. The territories are defended mainly against other magpies. Young from a previous year are part of these social groups, which may contain four or more mature birds. The young are fed for many months after leaving the nest. Their plaintive, begging call is well known.

The parents and others will feed the young on all manner of insects, including Christmas beetles. Large caterpillars and grubs are taken from lawns and cultivated land. Their ability to find curl-grubs under the grass is uncanny. They swoop down on the newly turned soil in your vegetable garden as soon as you turn your back. Feeding of the insistent young continues until dusk. The magpies' role in controlling pests is very worthwhile. Remember this when you are swooped on by an irate magpie, clicking its beak as it skims past your head.

There is no doubt that your garden belongs to the magpies as much as it belongs to you. For most of the year they are friendly and not at all aggressive. The immature birds are especially curious, and will come close while investigating your suburban activities. They will play with the things that take their fancy. If you have reared an abandoned young bird, it will help you in your tasks, tugging at a builder's line and playing with any tools left lying around. These tame birds are wonderful company on the early morning tasks around the garden.

The mistletoe-bird and the pardalote are two similar birds with quite different habits. Both may be seen on grazing properties and in suburban gardens. The striated pardalote is a tiny bird which feeds in small flocks in the outer foliage of eucalypts, searching the leaves and twigs for scale insects, thrips, lerps, moths, caterpillars, small beetles and spiders. In towns they frequently search under the window lintels for spiders. Spotted pardalotes nest in narrow burrows excavated in the banks of creeks and ponds; in the absence of such wet places, they will nest in a tiny knothole low down in the trunk of a dead tree. Their coming and going from these nesting places is not interrupted by your presence. Nest boxes under the eaves of a house, with a small entrance hole covered by a hollow log, will be used year after year by striated pardalotes and the number of these birds in your garden will increase.

If you have ever wondered how mistletoe is spread from tree to tree, you have not recognised the mistletoe-bird feeding on the fruits of mistletoes in gardens on the edge of town. This species has a rudimentary gizzard, and the alimentary tract is a uniform duct that enables mistletoe berries to pass through rapidly. The voided sticky berries may adhere to the branches of trees, where they germinate and penetrate the bark. The neatly shaped nest is suspended from a tree branch at any height above the ground, but in gardens they will nest in low bushes. The female builds the nest from plant debris cemented with cobwebs and covered with wattle flowers and lichen.

When feeding young, the male and female also take insects to the nestlings. The nests may be partly covered with the bodies of caterpillars.

Birds of many species are a source of joy and unending surprise in town and country gardens. There are many species that visit suburban gardens for short spells, so look forward to these passing visitors. If you list the species present in your garden as you identify them, you will be surprised at the length of the list by the end of a year. Noisy friarbirds, cuckoo-shrikes, finches, butcher-birds, cuckoos, mudlarks, kookaburras and galahs — your list of birds will never be complete. One day a new bird will appear fleetingly, such as a yellow-tailed black cockatoo which visits rarely in town, squawking loudly and tearing the bark and wood of eucalypts to feed on the borers deep in the wood of the trunk. The colourful eastern rosella and its brilliant cousin the crimson rosella reside on the tablelands. You can find them breeding in hollow stumps left standing in pastures. They destroy the seeds of thistles and other weeds. In gardens in town they are frequent visitors, attracted by the fruits of *Pyracantha* and other berry trees. Surprisingly, they also feed on insects. In many towns ibises and plovers (called lapwings now) will visit school playing fields. If there is water in a pond, many of the waterbirds will drop in unexpectedly from time to time. Honeyeaters and spinebills are attracted to nectar from flowering trees and shrubs, and as they feed their brush tongues are inserted into tubular flower corollas. Besides being useful pollinators, most honeyeaters also feed on fruits and insects. Several species of honeyeater and the eastern spinebill are residents of the tablelands. These colourful birds have a variety of songs, all musical.

The changing seasons bring changing bird populations. Why? The answer is usually the changing food supply. This raises another matter. The consequences of frequent and widespread use of insecticides are that many gardens are robbed of the food that attracts insectivorous birds. In a suburban garden in which vegetables, fruit and ornamentals are grown for home use and pleasure, there is little or no need for insecticides. Advice given to gardeners often ignores the value of biological control of pests. Many pest control agencies engender a fear that you will suffer a build-up of pests if you do not use preventive insecticides. Where there is a good balance between pests and predators this is simply not true, and there is no doubt that many times more pesticides are used in gardens than can be justified.

Photograph 6
Kookaburras, which belong to the kingfisher family, are well equipped to deal with snakes, lizards and grubs.

You may each choose for yourself the kind of garden you want. Whatever choice you make will be evident to birds by the presence or absence of the insects on which so many beautiful species of birds feed. Remember that they demand a continuous supply of food, without any lean periods. When you spray at the first sight of insects you are destroying the birds' food supply. Try instead a philosophy of wait and see. You will be surprised by the results. Many of your pest problems will just go away or will require only the mildest insecticides. A spray with soapy water is a useful deterrent to soft-bodied insects like aphids. Even a jet of water from a hose will knock insects to the ground, where lizards, spiders and other predators may find them before they climb up again. Stinkbugs and beetles can be kept in check by shaking them into a can. Plants growing vigorously, with adequate fertiliser, composted to loosen the soil, and with plenty of water are remarkably resistant to insects. Learn to recognise not only the birds, but all the other insect predators in gardens. Together these biological controls may keep pests at a tolerable level.

Count the birds in your garden. We have been listing here the species in suburban gardens, but the **number of individuals of each species** is a new dimension. When food is plentiful in your garden you will attract birds from around about. Birds will travel surprisingly far in search of food, and they will learn which gardens provide good pickings. After a few years of sensible gardening, you will be surprised and delighted by the increase in numbers of birds you had imagined were quite rare.

Just remember that all birds need a place to nest and plenty of insects in spring to feed their chicks — and nearly all birds in gardens and on farms feed their young on insects, which makes them good pest controllers.

8

Territory: An Ordered Society

If one bird were to be singled out above all others as the most useful in the control of pasture pests in Australia, it would be the magpie. Yet how many graziers have taken steps to encourage magpies onto their properties? To many townspeople magpies are a nuisance because they threaten (but seldom harm) humans for a few weeks in the year, when young emerge from the nest. Let us bring together a few facts about the social behaviour of these birds, as an example of a native species that can be encouraged to play a role in agricultural production and also adapt itself to suburban gardens. If we are not aware of the biology of such native species we can unwittingly destroy their habitat, as we are now doing to the magpie's habitat.

Magpies, *Gymnorhina tibicen*, occur in Australia and New Guinea and they have been introduced to New Zealand. There are three main subspecies — the black-backed, white-backed and western magpies — which interbreed freely. All are called the Australian magpie. Magpies are strongly territorial. A group, usually two to three adults and a few juveniles, permanently occupies a territory that varies in size from one to 15 ha or more and has fixed boundaries. These birds remain in their territory all day throughout the year and defend it vigorously against other magpies, but allow other species of birds to nest and feed. Ideal territories contain clumps of tall eucalypts and a nest is constructed by the female near the top of a tall tree. Although there may be many suitable trees in a territory, no female magpie other than the dominant female of the group may nest in them. Successful breeding of magpies takes place only in such territories.

As clearing of timber has progressed, many thousands of hectares have been left with insufficient remaining trees for breeding groups of magpies. Juvenile birds are ejected from the territories after one or two years, and when they cannot find a territory with suitable trees they remain non-breeding birds. Robert Carrick has described the social grouping of these birds. Apart from the **permanent** groups, there are **flock birds** that are non-territorial; a flock may number up to a hundred birds of all ages and both sexes. They feed in open pasture and roost in trees, but they do not attempt to breed. **Open groups** feed in treeless pasture, and fly in different directions to roost at night. From these roosts, which may be many kilometres from feeding grounds, they fly each day to regroup on pasture. They make no attempt to breed. **Mobile groups** also commute between open pasture and distant trees, in which they roost and attempt to breed. But breeding is not successful, the young dying at an early age. One other social group is common in areas with inadequate tree cover, where a **marginal group** occupies a territory with an isolated tree or even a telegraph pole in which a nest is built. Eggs are laid and hatched out, but the chicks seldom survive. These marginal and non-breeding social groups can be easily recognised because they do not contain first-year juveniles.

In magpies, the reproductive rate is usually far below the potential of the species, due to high egg and chick losses and in particular the large percentage of non-breeding females. Was there ever an ecological imbalance that could be laid more squarely at each property-owner's door? Here is a sedentary species, a bird that lives its whole life in an area of about 10 ha, so that one farm may have ten or perhaps 100 permanent territorial groups. Food is abundant as magpies feed on a range of insects in pastures, including underground ones like scarabs. During the year the food intake varies to take advantage of the commonest insect species. In the spring, magpies live largely on scarab grubs found under pasture or lawns. Robert Floyd found that magpies can hear the large grubs in the soil, and they pick them out unerringly by pecking at the exact spot. A group of students (Maureen Filmer, Des Foley, Dave Piggot, Lorna Sleep and Mick Thomas) found in a study at the University of New England that at this time of year the average daily catch of scarab grubs by each magpie was 42 grubs with a total weight of 35 g. Wilhelmus Vestjens and Robert Carrick have calculated the **hourly** intake of stick insects, moths and worms by magpies in captivity to be 6 g per bird.

A study of the stomach contents of 1319 magpies examined over eleven years in the southern tablelands by Vestjens and Carrick gives a clear picture of the food preferences of magpies. Ground weevils and ants were the most common food. Scarabs, caterpillars, grasshoppers, spiders and earthworms were present in about half the birds. Shield bugs, ground beetles and crickets were found in about a quarter of the birds. A wide range of insects made up the balance, but hardly any frogs or lizards or snails were found in the magpies, and very little plant material. There was a large variation in the food taken at different times of the year. For example, Christmas beetles (contrary to popular belief) and grasshoppers are taken by a large percentage of magpies in summer, while beetle larvae and earthworms are taken mostly in winter and spring. Weevils and ants are taken throughout the year.

No doubt magpies in pastoral regions in eastern Australia have abundant sources of food available to them. However, on pastures which have been improved to the point where patches of eucalypt woodland are so scarce that magpies cannot find enough territory, lack of adequate breeding places is causing a reduction in the magpie population.

Here is the perfect biological control — a native bird that feeds on a variety of invertebrates including ants, weevils, scarabs, grasshoppers and caterpillars — a most adaptable bird that will find food even when pest numbers are low, and will switch between prey as pests increase. To encourage magpies, farmers should be aware of many aspects of their biology. Isolated trees do not allow successful breeding because the nestlings are exposed to predators while the parents are foraging. Nesting territory is always strongly defended. A clump of eucalypts may contain many trees suitable for nests, but only one bird will be allowed to nest in the territory, which may include all the trees in that woodland.

The melodious carolling of magpies is an announcement of ownership of a territory, a possession maintained throughout the year. Last autumn we witnessed a challenge to the ownership of our garden, or should we say magpie territory? A sustained carolling in a dead peppermint tree (which is a vantage point favoured by all kinds of visiting birds) was followed by a tussle on the lawn, which lasted ten or fifteen minutes. But it was no common brawl, rather a formal play-acting more ancient in its ritual than any Japanese theatre. After a few aggressive passes one bird would be intimidated and would lie on its back to signal defeat. Immediately the victor would cease its attack, perhaps touching the vanquished bird with its beak or claw. These bouts continued until the protagonists were satisfied with their status. (Simultaneously, a separate bout was in progress a few metres away, two hens or possibly juveniles going through the same stylised movements.) The challenge was repelled, and our magpies resumed the confident patrolling of their territory. We seemed to detect an extra pride in their song. A few small feathers on the grass were the only signs of the determined defence of their home territory.

The grazing property 'Saumarez', near Armidale, is an old established station with well improved pastures. Although large areas have been cleared of timber, there is an extensive area of woodland in one corner and occasional old eucalypt trees here and there. It is an ideal site to study the movements of magpies. In a preliminary study, students Kate Boyd and Rick Phipps marked out equal sized plots in pasture in lines at right angles to belts of woodland in which magpies had established breeding sites. The birds appeared to feed mostly 150–300 m from the trees, rather than near or under the trees. But later, when fledglings were being fed, the young and old birds spent most of their time feeding under the trees.

On a treeless pasture on 'Saumarez' flock birds were counted in plots of the same size, all 700 m or more from the nearest trees. Flock birds were feeding regularly in pasture here, with an average of four birds per hectare. These non-breeding magpies did not roost on fence posts or telegraph poles nearby, but sought shelter elsewhere. Flocks like this feed on only a part of a paddock at any one time, and the density recorded above may be misleading. On 'Saumarez' the density of territorial magpies was one bird on two hectares.

Magpies eat vast quantities of insects. As they are permanently resident on your property, you may count the magpies, count the insects eating your pastures, and then prepare a balance sheet. On 'Saumarez' an average square metre contained ten scarab grubs, 145 weevil larvae, 13 earthworms and 40 'others' (spiders, millipedes and so on). In any year the magpies on 'Saumarez' would have eaten most of the large scarab grubs, and probably a proportion of the weevils when adults emerged. In a year of high scarab numbers more magpies would be necessary to reduce the pest to an acceptable level, but more trees would be needed to provide more breeding places and territories. Flock magpies were more common than territorial magpies on 'Saumarez'. These non-breeding birds may have come every day from a wooded hillside on an adjacent property, as they would fly more than 8 km each day to search for food.

During the three months each spring when chicks and fledglings are being fed, all the magpies in a territory collect more insects than at any other time. By good fortune this coincides with the time when most scarab grubs are at their biggest and fattest. These large grubs under pasture and in lawns are the

PLATE 20A **PLATE 20B** **PLATE 20C**

Pasture scarab grubs parasitised by scoliid or hairy flower wasps: (A) egg, (B) larva and (C) pupa. Photos by Chris Haywood, courtesy NSW Agriculture.

PLATE 21

Scoliid or hairy flower wasps may be seen in large numbers in late summer, feeding on nectar. Eggs are laid on scarab larvae which are eventually killed by the parasite.

PLATE 22

Sphecoid wasps paralyse their prey, and then pack them into mud cells and lay an egg in the cell before sealing it off.

PLATE 23

Leaf-eating sawfly larvae are usually seen in masses on eucalypt trees and they have a cooperative defence strategy if disturbed.

PLATE 24

Victims of the pathogenic fungus *Entomophthora* typically climb up grass stems or twigs before succumbing.

PLATE 25
Mistletoe-birds build a purse-shaped nest of cobwebs and plant fibre.

PLATE 26
Magpies locate scarab grubs under lawns and pastures by hearing the noises made by the grubs as they burrow with their jaws.

PLATE 27
Young magpie begging for food.

PLATE 28
A praying mantis consumes a moth.

PLATE 29
Flower or crab spiders station themselves on a flower and wait motionless, with forelegs outstretched, to ambush luckless insects visiting the same blossom.

PLATE 30
Wolf spiders show a certain degree of parental care. This female, at the entrance of her burrow in a partly improved pasture, is carrying a brood of spiderlings on her back.

PLATE 31
Huntsman spiders are often referred to as 'tarantulas', although unrelated. They are well adapted to live in confined spaces.

main food of magpies in spring. The insistent pleading of fledglings drives the adults on to feed late into the evening, and again before it is truly light in the morning.

Magpies in flocks up to 100 may be attracted to cultivated land where grubs are exposed by ploughing or harrowing. Bob Wickson, a horticulturalist with New South Wales Agriculture, persuaded several potato growers to make use of birds to destroy scarab larvae in land to be planted in October. An early cultivation of land in June or July with a spring-tine harrow attracted many insectivorous birds. Early the next morning the farmer counted the birds still feeding on 10 ha of land: 100 magpies, 30 magpie-larks, 15 crows or ravens and an occasional kestrel were feeding on the grubs turned up. A flock of ibises stood in the next paddock, as if to say: 'There won't be much left for us'.

Strangely enough, territorial magpies do not chase other bird species from the territory which they defend so vigorously against other magpies. Magpie-larks, themselves strongly territorial, will nest in the same territory, especially if there is water nearby. Their mud nests are attached like a small bowl to a bare horizontal branch of a tall tree. Crows and ravens may also share their nesting sites with magpies. These much maligned birds are mostly insectivorous, although they occasionally rob nests or scavenge on dead animals. It would be a good thing if the attitude of graziers towards them was changed, because they are most useful birds.

Veronica Parry has written a delightful book on the laughing kookaburra, which has a social behaviour that is almost human. Its territories are small compared with those of the magpie, and the boundaries are changed from time to time as the flocks compromise with one another, smaller flocks conceding territory to larger. During a two-year study in Victoria, Parry found that the size of kookaburra territories was determined by the number of birds in each family group. Only one mated pair occurs in a group, together with up to four juveniles from previous years, which do not breed but assist in incubating and feeding the chicks. In this study site the size of each territory was correlated with the number of birds in residence, large enough to provide 1.2 ha for each adult and juvenile bird. The smallest territory was 2.4 ha and the largest 7.2 ha.

Adjustments to kookaburra territories are made just prior to the breeding season, and adjacent family groups in which the numbers have increased expand into the territory of groups who have decreased because of less successful breeding, deaths and departures. The size of 1.2 ha per bird seems to be related to the maximum area that each bird can defend. Figure 16 illustrates these territorial compromises.

Kookaburra territories must contain a permanent food supply, at least one hole in a dead tree large enough to nest in, and an area of dense woodland for a roosting site and refuge for fledglings.

Kookaburras communicate by distinctive sounds and gestures, which are described in Parry's book. If you learn these, you will know when they are frightened or aggressive, welcoming or seductive. Apart from their beauty and interest, kookaburras are valuable pest controllers. Insects constitute 32 per cent of their diet; lizards and snakes 35 per cent; earthworms 15 per cent; yabbies 8 per cent; human handouts 7 per cent; and rodents 1 per cent.

Many birds defend their feeding and nesting territory against their own kind, while allowing other species to nest and feed. Hugh Ford, in his book *Ecology of birds: an Australian perspective*, describes how numerous species can coexist by taking different foods, or by foraging at different sites or in different ways. Because the diets are different and the species feed in different strata (see fig. 17) there is no conflict for food.

On the other hand, some birds are colonial, feed in small or large flocks and nest close together. Thornbills feed in small flocks, often together with other species. Noisy miners maintain colonial territories which they defend aggressively from other species. In these territories smaller birds are frightened away, and in gardens and on farms noisy miners (or soldier birds as they are appropriately called) should not be encouraged by feeding with bread and sugar or honey, or many of the useful insect-eaters will disappear.

An understanding of the territorial behaviour of birds enables landholders to increase the number and variety of birds on their properties. Food throughout the year is an essential part of territory. A place to nest, whether it be a living tree for magpies, a dead tree or a hollow limb in an old tree for kookaburras, or shrubs and tall grasses for smaller birds, is also vital. This is a reminder that birds and other animals have very specific habitat requirements. We would do well to see the world that we control in our urban and rural developments from their point of view, not just because they are beneficial consumers of pests, but also for the pleasure of their company.

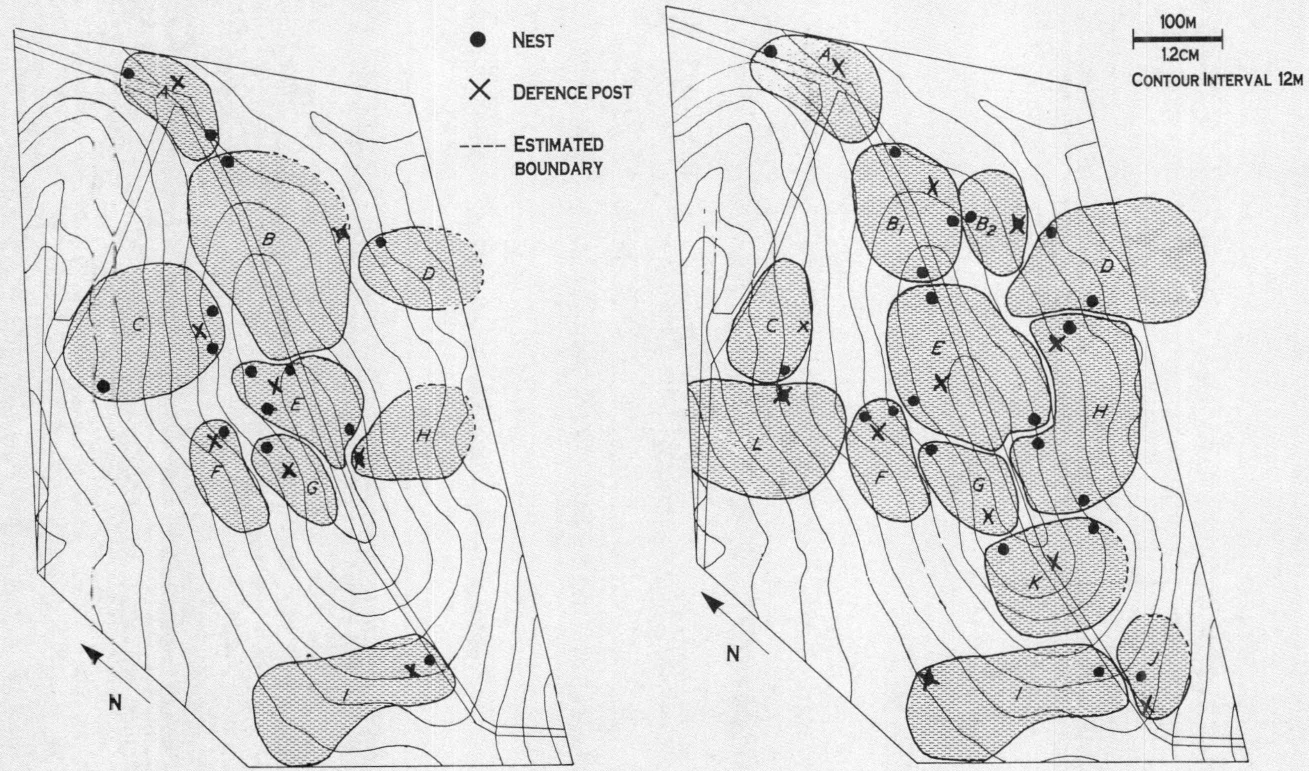

FIGURE 16

The territories of kookaburra groups in two successive years to allow for the changes in the size of the groups. From Parry (1970), by courtesy of Lansdowne Press.

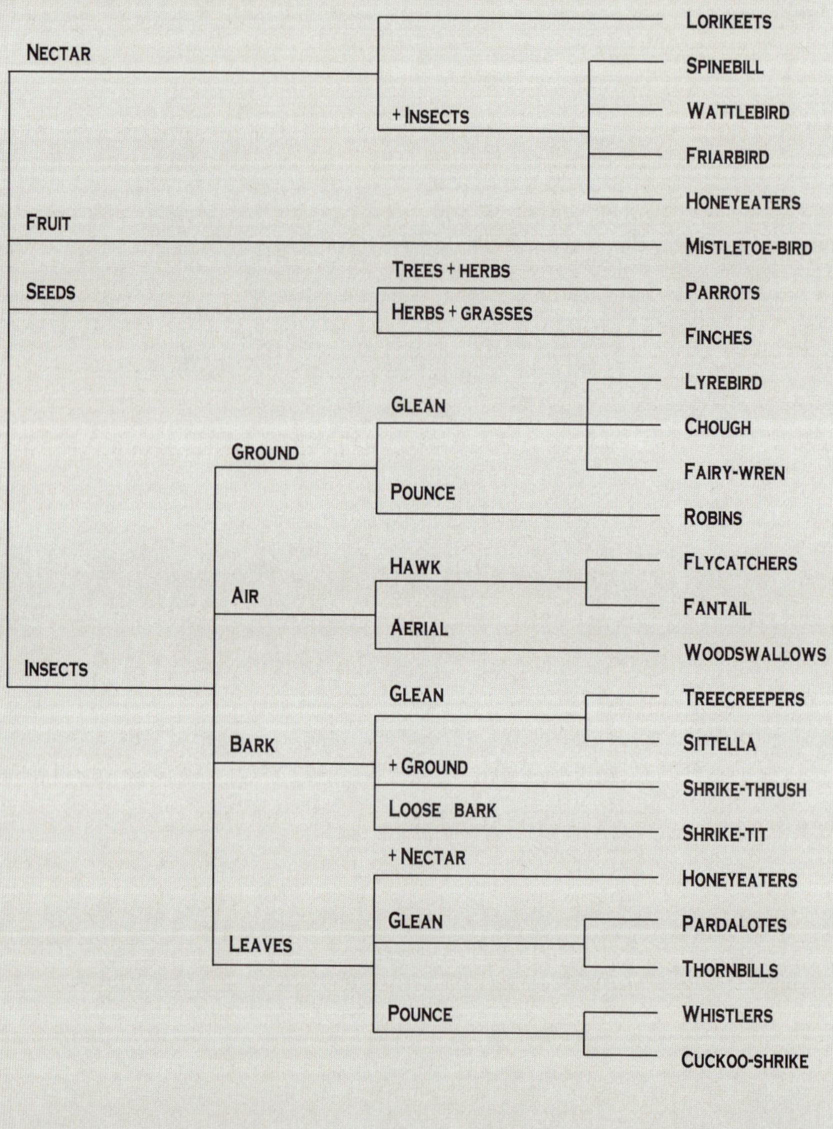

FIGURE 17

Partitioning of the bird assemblage of eucalypt forest in New South Wales by food, substrate and foraging method. From Recher and Holmes (1985) and Ford (1986), by courtesy of Surrey Beatty & Sons.

9

Spiders' Webs

One autumn we spent a night camped in the bush. A heavy mist came down next morning and the droplets of dew settled like jewels on the webs of spiders. We have all at some time admired an individual web reflecting the sun's rays. But on this morning we looked down the slope of a tree-covered hill, and against the sun rising like a pale moon in the heavy mist were thousands and thousands of webs. Long strands were stretched between the tall stringybark trees. Each blackthorn bush had webs less than a hand's span apart, so that each bush had 200–300 separate webs. On the grasses the webs were stretched between the flowerstalks and on the dry seedheads. Tiny webs had been spun between the lower leaves of the closely grazed grasses, and on the ground there were webs lining the burrows of other species.

When we looked away from the sun there were no webs to be seen, because the light was not reflected. This was an uncanny sensation; half the world (to the east) was barricaded with spiders' webs, while the other half (to the west) was open. No moth or fly or beetle could move in a straight line for 10 m without risk of being snared in a web. We realised then, on that misty hillside, that the value of spiders as predators is vastly underrated. The sheer weight of numbers in that woodland was a compelling argument for the recognition of spiders as a means of biological control of insect nuisances. Furthermore, many spiders are nocturnal, so daytime counts will inevitably underestimate their number.

In the months that followed we wondered, foolishly, why all these spiders dwindled in number in summer. When we read the literature on spiders, we learned that their numbers actually increase in summer! Because there was no heavy mist we were overlooking their presence. Then a week before Christmas, at home, there was again a heavy mist through which the rising sun shone weakly. The spiders' webs were sagging with the weight of dew drops. On a *Grevillea* bush the webs were about 10 cm apart. Large webs were stretched between the planks of a sliprail fence. A hole in the fence was filled with one huge web.

Since those misty mornings, when the webs were so conspicuous, our interest in spiders has grown daily. Recently, while photographing a pure white flower spider which was waiting, with forelegs outstretched in readiness, for a fly or bee to land on its chosen daisy, we saw another odd-looking spider. It resembled a small scorpion with a long tapered 'tail' and, taking a guess at the common name, we hurried to look it up in one of our books on Australian spiders. Sure enough, it was indeed a scorpion-tailed spider and we felt fortunate to have such an unusual resident. The mandatory photos were taken of the spider and her neat little web, but in the process yet another spider was spotted. This time a more robust orb-weaver sitting resolutely in her beautifully constructed silken trap, already strewn with the cadavers of past victims. She was like a queen reigning over her sphere of influence, with its borders marked in lines of silk. The analogy is not inappropriate, for the female spider is really quite imperious. Male orb-weavers pale into insignificance in the female's presence, usually being very much smaller and taking no part in the construction of the web. During courtship they must approach the female with great care or run the risk of being consumed. Her majesty, like all good monarchs, is well informed of the state of her land, with any intrusions immediately relayed to her by vibrations of the web. She is quick to investigate — it might be a trapped fly from which to suck a meal or even a cautious male tweaking at her web to attract attention like a nervous suitor at the palace.

Once the eye becomes trained to spider-spotting, they seem to pop up everywhere. We found a dozen or so spiders, belonging to three or four species, in that one small bed of daisies. Flies and beetles attracted by the scent of the blooms gambled with death as they moved from flower to flower, each time running the risk of being grabbed by a flower spider or blundering into a web. There were also green praying mantids lurking amongst the stalks and flowers. Instinctively they lurched and leaned towards grasshoppers passing close by, ready for that last explosive lunge, spiny raptorial forelegs closing on the prey. First grace, then the meal!

Most gardens are alive with a variety of predacious invertebrates but the spiders are especially notable and mostly harmless to humans. When the weather is favourable you can often observe the deft skill of the orb-weaving spider as she builds her snare. The legs are used to space the silken strands correctly and to help draw the silk from the spinnerets on the abdomen. Some spiders hang upside-down in the centre of the web all day; others take refuge in nearby vegetation during the day and take up station on the web in the evening. Resourceful spiders of the genus *Phonognatha* suspend a curled leaf in their web and hide inside. When the web is rebuilt close by, the leaf goes too.

Not all spiders construct orb webs. The Theridiidae, to which the infamous red-backs belong, weave untidy cobwebs as do the daddy-long-legs that are so fond of the comforts of human dwellings. Others build platform webs or sheet webs. To our untrained eyes, all spiders build 'cobwebs' and they often suffer an undignified end as a result of our neat and tidy ways. Often the sinister warnings in pesticide advertisements are only too effective and the householder rushes for an aerosol spray because of a cobweb on the fence or a spider in the garden. We all know of the unpleasant reputation of red-back spiders and funnel-webs, but it is surely over-reacting to annihilate anything that crawls in the garden on this basis, or because the webs look a little untidy in the eyes of the beholder. The continued use of poisons against spiders reduces predation pressure on garden pests, and the end result may be even greater use of pesticides to control insect pests which previously would have been largely controlled by spiders and other natural enemies. Most spiders found in gardens and homes are probably less harmful to man than the poisons used against them.

Lycosids or wolf spiders often burrow in grassland and emerge to actively hunt for insects to eat. Laboratory feeding trials in which wolf spiders were provided with black field cricket nymphs as prey indicated that the hunters can consume the equivalent of almost 10 per cent of their own body weight per day. When immature spiders were given very small cricket nymphs, they consumed on average 14 of these per day over ten days. They look fearsome, but Australian lycosids are not regarded as dangerous. Wolf spiders attach the egg cocoon to their abdomen. It looks like a miniature ping-pong ball as it is aired and sunned at the burrow entrance. When the spiderlings emerge they clamber onto the mother's back and jockey for positions, even crawling over her eyes.

Many spider species do not build webs at all, although small amounts of silk are produced to act as a lifeline when the spider jumps, or for protection

PHOTOGRAPH 7

A spider's jewels. Early morning mist emphasises the perfection of an orb web spanning the gap between two weeds.

of eggs, or to aid in dispersal. The delightful and perfectly harmless flower spiders have been mentioned but perhaps even more charming are the jumping spiders. They can indeed jump and their quick movements often attract one's attention on a paling fence or window-sill. Two of their eight eyes are very large and this binocular visage coupled with sudden craning movements of the head and general alertness are most monkey-like. Their antics are always endearing, especially under a magnifying glass or microscope.

In the bush the diversity of the spider fauna is an education. Each species seems to have found itself a particular niche and a special way of life. It is as if the total living space, the habitat, has been carefully divided up to allow as many different types as possible to make a reasonable living. It is a well-respected ecological principle that this partitioning of habitats **does** occur as a result of evolutionary processes. Some spiders, most notably the huntsmen, set up home under the bark of gum trees. Others, such as the New England funnel-web, reside amongst debris in the forks of trees. Orb-weavers span their webs between grass stems or branches while wolf spiders are usually ground dwellers and generally have a burrow of specific design close to their hunting ground. Trap-door spiders, as a rule, build a neatly fitting door to their burrows. The homeless lynx spiders have no permanent retreat at all. Nursery-web spiders prefer damp situations and even hunt on the surface of ponds or streams. Many species share a liking for a solid rock roof, amongst them a crafty spider (*Supunna* sp.) which seems to mimic a wasp. It moves in waspish darts and the pedipalps and front pair of legs, which are yellow, move feverishly when the spider is upset, in much the same manner as a wasp's antennae. Other spider families are abundant in leaf litter.

Whatever their particular interest, Australian araneologists live in a veritable Garden of Eden. But so do we all and we can gain much pleasure, without the responsibilities of a research scientist, from simply observing the antics of these intriguing eight-legged, eight-eyed mini-monsters.

This chapter began with a description of thousands of webs seen one misty morning. We became aware at the same time that without undergrowth or trees or tall grasses there would be no foothold for many of these webs. Similarly, Barbara York Main, in her inspirational book on spiders, has described the detrimental effects of ploughing and trampling of stock on burrowing spiders, which find the direct disturbance too severe or the soil too compacted. Unlike many agents of biological control of pests, such as the *Cactoblastis* moth, the spider fauna is indigenous and therefore doesn't have to be imported at great expense and after much research. Recognising the importance and value of the spider fauna as an agent of pest control we again come back to the need for relatively undisturbed areas of woodland, forest or even grassland where spider populations can be maintained. Secondly, both large-scale and small-scale use of pesticides which are harmful to spiders should be discouraged. The combined insect kill by many thousands of spiders in a pasture or crop, effortlessly and free of charge, far surpasses the ability of man to achieve effective, long-term control of insect pests by spreading costly toxic chemicals. Problems of insect pest resistance are also avoided.

If we recognise the worth of long-term and widespread biological control of agricultural and garden pests by indigenous predators we would be well on the way to achieving one of the most advanced, sustainable and economically viable rural economies in the world.

10

Of Rocks and Logs

Its head tilted enquiringly — an eye glinting in appraisal. Who was this rude intruder seated on its log? The stranger, a human, made a slight movement and the lizard scuttled back into its log only to emerge a few moments later, sheer curiosity overcoming fear. It was a small skink or penny lizard and its interest was equal only to mine.

Maybe it's their primordial aura, but dark quick skinks, coarsely-clothed Jacky lizards, bearded dragons and cumbersome bluetongues, like all reptiles, never cease to fascinate us. Who can help but admire the smooth-skinned blue-tongue with its face-pulling threats? We probably remember most lizards by their characteristic defence strategies. The proud stance of the bearded dragon attracts us too close and the transformation is startling. Doubling its size in an instant, with its yellow mouth gaping and beard extended, the lizard usually discourages further interference.

A related dragon, commonly called the Jacky lizard, is also a master of camouflage and flees only when confrontation seems imminent. Juveniles are caricatures of their parents with globular eyes and oversized feet. Animated, real-life cartoon characters performing on makeshift stages of rock or sunbathing decadently on warm logs.

The legless lizards closely resemble small snakes, having no fore limbs and only rudimentary hind limbs, but their method of escaping predators is ingenious. If caught by a bird, a legless lizard effects an Houdini-like escape by amputating the tip of its tail — literally sacrificing its tail to save its skin. The bird is left with a still-wriggling tail to occupy its attention while the lizard steals away to regenerate a new one. Skinks and geckos also employ this escape mechanism.

Lizards, being reptiles, are cold-blooded or poikilothermic, relying on external sources of heat. They have become adept at regulating their body temperature by basking in the sun when cool and changing their posture or moving to shade when too hot. To their detriment, many reptiles seek the

PHOTOGRAPH 8

Fallen logs and dead trees should be left in woodlots for birds to nest in and as shelter for other animals. They will decompose in time.

PHOTOGRAPH 9

Perfectly harmless, the three-toed skink is a common inhabitant of garden rockeries. It is a beautiful shiny lizard with a surprisingly orange belly.

warmth of bitumen roads and often meet an untimely end. This behavioural thermoregulation is augmented by colour changes in some lizard species. Dark colouration increases absorption of solar energy while light tones increase reflection. These colour changes are usually brought about by contraction or expansion of special pigment cells which are found in the skin. These cells contain a dark pigment and when contracted give the lizard a lighter colouration.

Many lizards will pant when their body temperature exceeds a temperature known as the panting threshold. Panting allows rapid heat loss by evaporative cooling. Some species will pant after exertion. For example, after fleeing a predator the Jacky lizard will be seen to pant and a herpetologist would say that it is repaying an 'oxygen debt'. The ability to pant is most probably a dual adaptation improving the animal's control of body temperature and also allowing efficient ventilation of the lungs in emergencies.

Lizards commonly encountered in suitable farmland and gardens, such as the bearded dragon, the Jacky lizard, the small skinks and the blue-tongued lizard feed on a wide variety of organisms. Some lizards, in turn, fall prey to other animals higher in the food chain. Many snakes, particularly the red-bellied black snakes, occur widely and include skinks and other small lizards in their diet.

Although most lizards are fairly opportunistic feeders, some general differences in food preferences exist between species. Small common skinks feed predominantly on tiny insects such as ants, springtails, spiders, beetles and termites, or on the eggs of insects, for which they forage amongst leaf litter and around rocks and logs. The larger dragons hunt voraciously for worms, mice and larger insects such as grasshoppers and beetles, being remarkably fast on their long legs when warmed by the sun. Many dragons also feed on small skinks. Blue-tongues have catholic tastes, being partial to both animal and vegetable matter. Gardeners often encourage blue-tongues by feeding them scraps of banana and in return they help to control snails.

Young reptiles are generally left to fend for themselves with no parental care and juvenile mortality must be high. Most of the lizards mentioned lay soft-shelled, leathery eggs in a hole dug by the female. The eggs are then covered with soil and left to incubate until the hatchlings emerge. However, some reptiles, such as the blue-tongued lizard and the red-bellied black snake, give birth to live young. Young eastern brown snakes are often striped and are easily confused with the eastern tiger snake. Many of the so-called elapid, or front-fanged, snakes are highly venomous even as youngsters. However, it is fortunate that most snakes are amazingly good-natured, shy and retiring creatures if left well alone. Admire their effortless grace from a distance and you will soon realise why these too are protected fauna. Their elegance is irrefutable.

Snakes can survive long periods without food, but by the same token a brown snake will quite easily devour a dozen or so mice in a day if given the chance. Even large prey are no trouble to a snake. A detachable lower jaw is pretty handy when your eyes are bigger than your mouth!

For the purpose of controlling insect pests or rodents, then, lizards and snakes are useful and beneficial. We would be sad if they declined or even vanished altogether through destruction of their habitat or thoughtless killing. The charming skinks, the military dragons and even the often heart-stopping snakes all serve to make life more interesting and our environment more diversified and stimulating. Even if we only occasionally see a skink on our back doorstep or catch a glimpse of the common blue-tongue at the roadside, we are the richer. Sentimental maybe — yet we are quick to complain of the sheer boredom, the never-ending monotony, of a high-rise city apartment. We still look forward to a weekend in the bush or a Sunday afternoon picnic with the kids. We may have to think a while, if asked why we enjoy these outings so much, but surely it is the trees, the grass and streams, the rocks and logs and the things that live in them, that make time in the bush unforgettable. Who would enjoy a barbecue on an acre of lifeless concrete?

Just as we need a roof over our head reptiles need their long grass, rocks, dead trees and logs. For example, in a 1981 study of four species, Bronwyn Crome found 210 lizards per hectare in a native grassland with scattered trees and 1510 per hectare in grassland with dead trees and litter and logs on the ground. We should think twice before clearing up dead trees, fallen logs and boulders — they are the habitat without which lizards and the like would not survive. Rocks, litter and logs are not only needed for shelter. They also harbour the beetles, ants, cockroaches and other invertebrates on which many reptiles feed, so they may be regarded as the pantry as well as the living room for lizards and snakes.

Other animals also rely heavily on rocks and logs for a living. That all-Australian creature the echidna, or spiny anteater, has strong flattened claws for overturning rocks and digging up ants and other insects that are then devoured with a sticky tongue about a foot long. During winter echidnas hibernate, often under a log or inside a hollow one, until the next spring when they can be seen ambling about in search of food. At this time queen ants, rich in energy, are especially sought after and large echidna-sized holes appear in many ant mounds. Otherwise echidnas are rather independent, carrying about a fine armoury of spines to deter predators. An ability to dig straight downwards is a useful ploy against introduced foxes and mischievous dogs.

Sugar gliders are common in eastern Australia in forest and woodland, especially where patches of acacias remain and tree hollows are available for shelter. Gliders feed on nectar, gum exuded by acacias and the sap of eucalypts, and also on insects found on eucalypts. They are a useful aid in pest control and their diet depends on seasonal availability. On farms the retention or replanting of corridors linking bush remnants with forest and woodland will encourage sugar gliders, and they are locally common (up to ten per hectare) where hollow trees remain (see Graeme Suckling 1983).

Bats belonging to several families are common on farms and useful predators, feeding on a wide variety of insects. Gould's wattled bat is abundant throughout Australia in woodland, shrubland and urban areas. It roosts in dead trees. A study in Victoria showed that its food consists mainly of scarab beetles, field crickets, moths and caterpillars (see Joan Dixon 1983). Mastiff-bats, long-eared bats and broad-nosed bats are also widespread insect eaters which roost in hollow trees and under loose bark (see Greg Richards 1983).

Perhaps surprisingly, scorpions are quite abundant in many parts of Australia. They may often be found under rocks where the environment is equable. Scorpions are formidable predators and feed on many other invertebrates. The victim is held with strong claw-like pincers and injected with venom from a sting on the scorpion's tail. Scorpions show parental care of their young and occasionally one may find an adult under a rock with her youngsters riding piggy-back.

Like a rainforest on a smaller scale, a rock or log is habitat for a host of interacting organisms. This is not to say that these so-called microhabitats are independent of the parent ecosystem, such as a forest, but the shelter they provide is essential to maintain diverse communities of plants and animals. Species diversity depends largely on the variety of microhabitats within an ecosystem.

Having recognised the need for these various microhabitats, and being aware of man's almost unlimited capacity to modify and manipulate whole ecosystems, there is a very great need to ensure that a variety of habitat types is preserved. We should develop a consciously responsible attitude towards the environment which we ourselves must inhabit, and over which we exert such autocratic control. Spare a thought for all those small inhabitants of long grass, rocks, fallen logs and standing dead trees!

11

DISAPPEARING HABITATS

Travelling through the northern tablelands along the New England Highway, one is struck by the number of dead trees. First awareness comes after climbing the Moonbi Range, and then increasingly the feeling of exhilaration in the cool freshness of the high country is tempered with disappointment that so many trees in the roadside paddocks are dead. Many of these trees have been ringbarked. The bleached trunks, stripped of their bark by the weather over nearly 100 summers and winters, with only the stoutest limbs proof against the tearing wind, stand like sentinels guarding the endeavours of the early settlers. The lush green pastures, sprinkled with the clean white flowers of clover in spring and autumn, and the fat Hereford cattle and fine-woolled Merino sheep are the proud achievements of these settlers.

However, you will realise before you reach Uralla that many of the handsome gums and stringybarks that were left standing by the graziers for shelter, and because they made a pleasing landscape of scattered trees and patches of woodland on the ridges, are dying. Their upper branches are leafless, and the green shoots on the lower branches look unhealthy. Many trees are already dead, the finest branchlets intact, showing that it is not long since they died. Our first impressions of the district, in the early sixties when we travelled widely through New England, were of the only dead trees being the ringbarked ones; only rarely did we see other dead trees, and they were usually old. We have talked to many people: graziers and scientists and the older residents who remember what has been going on. We have seen the rising concern of the landowners and the townspeople at the uncontrollable death of these trees. To understand what is happening we must trace the history of white settlement of these tablelands. A similar history of agricultural development has occurred in other parts of Australia, and rural tree decline is widespread in all States. We are nearing the end of our book, and must try to answer the question: 'Where are we headed?'.

Surprisingly few people appreciate the viewpoint of the early settlers, and the uncertainty of their daily life in a strange new environment. They had many decisions to make, without the advice of agriculturalists or even experienced neighbours. The tablelands were covered with an open woodland of tall eucalypts; a sparse understorey of flowering shrubs like wattles and blackthorn; and a varied perennial grassland of kangaroo grass, wallaby grass, red-leg grass, lovegrass and snow grass. If you wish to transport yourself back 150 years in history, take the trouble to drive through the Eastwood State Forest only 10 km from Armidale, or walk through the Avondale State Forest near Rockvale. Although each is only a few hundred hectares, you will share the feelings of the early settlers who were enchanted by the tablelands. The trees are magnificent: tall yellow box as clean as a ship's mast, with a solid crown of tiny grey-green leaves; the stringybark with a slight buttress to anchor the soaring trunk, straight as a die; and the graceful red gum and manna gum, their bark smooth beneath the long ribbons of last year's bark peeling off. There are long vistas of yellow-green grass, pale in the cool shade of the widely-spaced trees. The ground is loose and friable beneath your feet, feathered by the leaf litter and fallen twigs. Lizards and spiders, beetles and moths — an infinite variety of wildlife, forming together a stable community. A hundred species of birds find home and food here; some in the tallest trees, searching in the foliage, creeping up the trunks to search in bark crevices, feeding on mistletoe; others nesting in shrubs and the hollows in old trees.

You could believe that the woodland, including all these species, would last forever, dynamic, the young replacing the old. A dead tree trunk, filled with termites, is a haven for echidnas and lizards. A eucalypt killed by borers is soon replaced by a sapling already there to take its place. The woodland seems indestructible. Before you read the rest of this chapter, ask yourself what you would have done in those times. Try also to feel the impact of subsequent events, so that you understand the actions of those men and women, and know how daily experiences shaped their views and the outlook of the children growing up in these surroundings.

There were restrictions on the cultivation of land, perhaps to protect the agriculture in the Hunter Valley. The tablelands were to be used as grazing properties. Cattle and sheep were herded from the earliest days, and throughout the year these were dependent upon native grassland, even in the harsh winters. While palatable enough in the spring and early summer, these grasses were of little value as grazing in the winter. Livestock tended to graze the more palatable species, and the other grasses were left until too dry to be eaten.

As in many other parts of the world, the dry grass was burned off in late winter when the grasses were dormant because of the cold weather. Eucalypts were not harmed by the fires, but much of the plant litter on the ground was burned away. However, the annual burn achieved its objective — the fresh green grass in the flush of spring was not obscured by dry, unpalatable leaves and stalks. This early growth each spring was awaited with anxiety by graziers who were always short of pasture in late winter. August was called 'starvation month'. Annual burning off is still practised in many parts of the world where native rangeland is grazed by cattle and sheep. It is a tool used by early agriculturalists, and used long before by nomads the world over, before and after the domestication of livestock. Foresters use controlled burning today to reduce the amount of flammable debris in woodland and forest. To them, the choice lies between regular, cool burns at chosen times, and destructive, accidental fires after an accumulation of dry timber and grass.

On the tablelands, burning of grassland in winter continued until about 1940 and later in some parts. Cattle and sheep numbers on the tablelands increased steadily until 1910, when the total livestock population levelled off, not to increase again until 1960. Fire, grazing and trampling made an appreciable impact on the soil and vegetation. We have no measure of these influences and we can only guess that most of the understorey shrubs were removed. To the birdlife there was a loss of habitat. The birds that nested in low shrubs probably declined; some insects may have increased because predators such as birds, lizards and spiders could not find a foothold. But these changes were small compared with the clearing of the woodland on a huge scale.

Imagine yourself a squatter on the northern tablelands, in possession of several thousand acres (no, not hectares in those days, 150 years ago). It is an uncertain possession, and you imagine (quite wrongly, as it turns out) that you have to show that you have improved your holding, or forfeit it. First you build a house of split logs, and almost at the same time sheep hurdles to yard the sheep at night. Mobile watch boxes are built for the shepherd sleeping with the sheep, who has to watch out for dingoes. Later, cattle yards are built. Two-handed saw, adze and axe are your only tools. Stringybark is much prized because it splits easily and does not warp, and the bark comes intact from the logs in large sheets for roofing. Yellow box is a fine timber, fine-grained and smooth when sawn, excellent for fence posts, slip rails and beams. Interior finishing, doors and furniture are made with cedar, which is close and plentiful. You clear the trees on your homesite, chosen because it overlooks the creek or a view of the hills. Huge trees are plentiful, 3 ft in diameter. Besides the timber they provide, this clearing is your protection against fire.

You now have valuable possessions, a home in a strange country. The years ahead will build up in your family an understanding, a measure of the new environment. You will feel the strength of the wind, the extreme of winter cold, the hot dusty weeks stretching out to rainless months beyond your expectations. One of your surprises will be the vigorous suckers and shoots that grow from every stump of the felled trees. Each clearing grows a thousand seedlings of stringybark, red gum and yellow box. Suckers and seedlings grow into saplings so close together that you cannot pass through on horseback. A fear rises in you that trees will soon reclaim your efforts. Your neighbours reinforce these fears by recounting their own experience. Imagine this being a topic of conversation in a chance meeting with friends in the little town of Walcha in the 1850s. From talk of two trees now growing where one grew before, you will fall to speculating whether the grass would grow better if all the trees were cleared from pasture. What a dream to take home that night!

The bush became oppressive to most landholders and the trees were cleared with a persistent zeal that can only be admired. You can still find the stumps of the early clearing done with axe and crosscut saw. The stumps have been shaved with an axe to prevent reshooting and these faceted trunks, covered now with lichen and moss, tell their story eloquently. But for the time being it was hard work in pursuit of an unattainable dream.

By 1860 the gold rush at Rocky River had petered out. The population of 5000 dwindled to 300, and the labourers (Chinese and European) displaced from the mines were willing to work on the properties, clearing timber. About the same time it was discovered that girdling the eucalypts would kill the tree and suckering was less than if the trees were felled. The bark was circled in two rings a hand's span apart, and the strip of bark peeled away easily. Only light strokes with an axe were needed. The pace of man's impact on the ecological balance quickened. Was there a feeling of triumphing over nature? We cannot find a description of the mood prevailing in those days. Nor do we know precisely the objective. Were labourers told to clear everything — the big trees and the saplings — leaving no eucalypts on the hillsides, the flats or even in creekbeds? There is no doubt that some followed this instruction exactly. Today in different circumstances we are amazed and saddened to find huge trees on the steep sides of a creek ringbarked. But we did not grow up threatened by the encroachment of these trees. The ringbarked trees did not grow shoots at the girdle, but new shoots arose from the base of the trunk. Usually one, two or three stems grew quickly from many of the rootstocks, and in about 20 years they were trees as thick as a man's thigh. Seedlings of eucalypts also grew into sizable trees. Many parts of New England were cleared not once, but three times between 1840 and 1940. We have spent hours examining these eloquent reminders of our past. We can read 100 years of history, of the settlers' hopes and frustrations in these tree trunks. Little has been written about the decisions made on clearing timber, and there is some guesswork in this reconstruction.

We will never know how the clearing of timber on the tablelands influenced the pasture. Peter Wright says that ringbarking heavily timbered country resulted in a spectacular, but short-lived, improvement in pasture quality. The native grassland was heavily stocked. After the Robertson Land Act threw the country open to selectors in 1861, closer settlement resulted in increased livestock numbers until 1910. It seems that the livestock by then had reached the maximum carrying capacity of native pastures. Further clearing of timber did not allow any increase in livestock. Burning off was still done each year, and regeneration of eucalypts was still a serious problem. For all the years between 1910 and 1950 the tablelands remained intractable. It seemed that the land had yielded up just so much, and could give no more. One to three acres was needed for each sheep. For 40 years there was no increase in the numbers of cattle and sheep. There was a livelihood here, but no fortunes were made.

World War One came and went, followed by the carefree twenties; the depression was endured; and the call to arms in World War Two was heeded.

Throughout these years the pride of ownership of the land did not falter. These settlers loved their land. They tried everything to improve their holdings. Fencing was erected to try rotational grazing. The grasses that grew so well in England and New Zealand were sown, and clover seed was scattered from the saddlebags of the stockmen. If these species germinated, they soon disappeared. The logs from clearing were burned on the ground, and the ringbarked trees and stumps were fired where they stood. We can see today the charred outlines of these stumps. The tidying up of grazing paddocks became a part of the ethos of those times.

Without being critical of these impacts of man, we must accept that the habitat of many animals was being ruthlessly destroyed. Birds that nested in hollows, like kookaburras, and rosellas that nested in hollow stumps were deprived of an essential requirement. Numbers of lizards, spiders, and all the beetles like carabids that live under the shelter of bark and fallen leaves, declined quite inevitably. To a degree the destruction of these habitats was deliberate. Rabbits were a plague, and the shelter provided by fallen logs was thought to enhance their survival. Foxes and dingoes were always a threat, and the hollow logs were believed to be favoured lairs. Blackberries threatened to take over cleared pastures. The common blackberry had been introduced by an orchardist in 1850 as cuttings from the United States. These had spread slowly at first, but the bushes now seemed to be everywhere; they grew to an alarming size, especially on the ashes of burned logs. Rabbits thrived under the protective cover. Many who were youngsters in the thirties will remember the hopeless task of cutting blackberries and sweetbriars. The New England Tablelands Noxious Plants County Council was constituted in 1947. They engaged contractors to spray with sodium chlorate, but it was ineffective and dangerous. One contractor was seriously burned using it. The story of successful weed control belongs to the era of the confident fifties.

In the fifties the dreams of the farsighted graziers began to come true. A few years earlier it had been realised that a deficiency of soil nutrients (mainly phosphate and nitrogen) was the main limitation to pasture improvement on the tablelands. By the simple expedient of aerial application of superphosphate fertiliser, followed by the broadcasting of clover and grass seed, evergreen pastures were achieved. Sheep and cattle numbers on the tablelands began to increase. There was talk of carrying four to eight sheep on each acre. Burning off was no longer necessary, and the struggle against regeneration of eucalypts became a thing of the past. In the years between 1955 and 1976 total livestock numbers more than doubled. There had been no increase in the area of land for grazing, but a doubling of the carrying capacity of the tablelands pastures. Some pastures were only partially improved, and native grasses were growing interspersed with clover. Some pastures had been completely converted to white clover and ryegrass, phalaris or cocksfoot.

In the new pastures, weeds were still a problem which was getting worse. Blackberry and sweetbriar had spread alarmingly. The newly established Noxious Plants County Council appointed full-time weed inspectors with spray equipment. There was plenty for them to do. The brush-killer 2,4,5-T became available around 1949 and inspired new confidence. The roadsides were sprayed assiduously, and properties were sprayed on contract or by landholders themselves. The Pastures Protection Boards maintained staff trained in pest and weed control, and the travelling stock reserves and considerable areas adjoining the roads were sprayed. In the improved paddocks a new weed appeared. Spear thistle seed had been introduced inadvertently with pasture seed from New Zealand, and the saffron thistle appeared like magic in cultivated land. But the mood of confidence was sustained. The hormone weedkillers were widely used; 2,4,5-T was solving the century-old problem of blackberries and 2,4-D was effective against thistles. When nodding thistle became a noxious weed, these controls were already at hand. Today a wide range of weedkillers is used for weed control throughout Australia.

Another problem that was worrying the grazier was the damage done to improved pastures by insect pests. In the late fifties and through the sixties there was mounting concern at the localised damage to the new green pastures by a wide variety of indigenous insects. *Oncopera* caterpillars came from their tunnels at night and fed voraciously on clover and grass, especially in the higher rainfall regions. Scarab larvae were found everywhere on the tablelands. Some adult scarabs, like Christmas beetles, were seen on the eucalypts during summer days. Other scarab adults could be seen at night also feeding on eucalypts. Crickets and grasshoppers all seemed to have their go at the improved pastures. An ecological study of pasture insects was being done in New England, and the commonsense view prevailed that insecticides should not be used on these pastures. The populations of all these native insects have settled down at economically acceptable levels which show no signs of progressive increase.

There was drought to contend with too. In 1964 the early spring rains were followed in 1965 by a dry summer and winter that devastated the temperate grasslands. Clover disappeared and the introduced grasses did not stand up well to the prolonged dry weather. Hand-feeding of stock was necessary, and as the expense of purchasing wheat and hay to keep stock alive mounted, each stock owner suffered the agonising decision: to keep or to reduce his herd? All learned the simple truth that the stock carrying capacity of the new pastures on the rather warm tablelands, subject to periodic dry spells, would be more moderate than early experiments had suggested. Cattle fared badly in the drought. While the sheep nibbled the stembases of grasses, and ate any organic matter on the ground, cattle were less adaptable. In their search for food, however, the cattle ate the outer bark of stringybark trees as high as they could reach. In their search for food when hungry, cattle and sheep will eat far more than clover and grass. The long-term effects of livestock on vegetation need careful consideration and we will return to this topic in a moment, because it is not only in a drought that domestic stock change the habitat of native species.

Many trees died in the 1964–66 drought. To anyone flying over the tablelands, the dead trees were conspicuous on the ridges. A very useful study of the effects of drought on eucalypts at this time was made by Errol Hoult. Young and old trees of many species of *Eucalyptus* were found to be dying in all topographic situations, but mainly on hilltops. But the drought passed and few people were concerned about the loss of eucalypts. Apiarists who watch the flowering of gums, boxes, stringybarks and ironbarks on the tablelands, moving their hives to catch the flow of honey in these trees, say that in the seventies there was little flowering of the eucalypts on the northern tablelands.

What happened in the seventies? Can it be that, only a decade or two after the dreams of the early settlers came true, those dreams were going sour? The pastures were green, the cattle were fat and the sheep were as placid as you could wish. Those who had built up a thriving pastoral industry should have had time to sit back and enjoy the beauty of their surroundings, but those trees

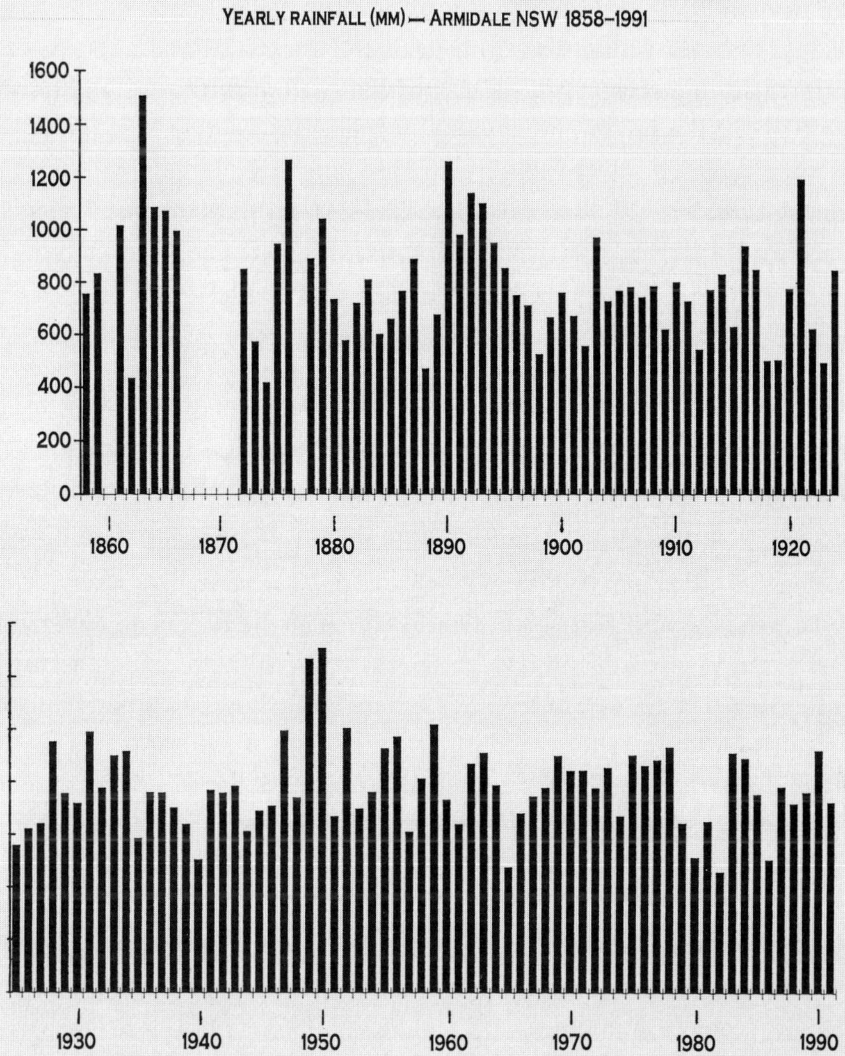

FIGURE 18

Fluctuation in rainfall is a powerful driving force in agricultural ecosystems.
Drawn by Peter Burr from data supplied by the Commonwealth Bureau of Meteorology.

that were left for shade were dying. Millions of eucalypts on the northern tablelands died in the decade 1970–79. Why the tablelands and not the slopes and plains? On a travelling stock reserve on the Rockvale Road near Armidale every tree died in just four years. On the hills nearby, young and old trees (stringybarks, red gum and yellow box) were dying. Dangar's Lagoon near Uralla is a picnic spot where visitors go to see the innumerable waterbirds — swans, coots, ducks and dabchicks. The birds were still there, but visitors asked: 'Why are all the trees dead or dying?'.

The concern was so great that the graziers and townspeople from throughout the tablelands launched an appeal for the New England Eucalypt Dieback Research Fund. A meeting in the Armidale town hall, one afternoon in July 1979, to launch this appeal was supported by 700 people. They were concerned about losing the character of the New England tablelands — the smell of the gum trees, the grey-green colour of the trees that outline the hills, the endless variety of eucalypts and native shrubs. A topography of hills and valleys without trees was unthinkable.

If you own land and work on it, would you wish the skyline to be stark and treeless? Would you sweat all day in the sun without the relief of passing under the shade of a grove of trees? What are your thoughts when you see a mob of cattle or flock of sheep too large for the shade of one spreading, twisted gum tree, drawn to it because the leafless branches of the other trees in the paddock cast no shade? When the chill westerly winds blow in winter and spring is there adequate shelter for newly shorn sheep and lambs? New England graziers gave a clear answer to these questions by their wholehearted support for the research fund.

All these people showed their deep concern. They were aware that their own environment was changing in a way they did not wish it to. They had lost control of the forces of nature. The face of their world was changing very fast, but could they halt this process that no-one understood? The newspapers, radio and television followed their dilemma and applauded their determination to save the natural beauty that was disappearing from their environment.

What will the pages of history record about this turning point?

12

Where are We Headed?

For 150 years the moist temperate regions of Australia have been developed with scant regard for the thousands of species of plants and animals that are part of this ecosystem (fig. 19). Some of the clearing for agriculture is recent, done with heavy machinery and herbicides instead of tedious hand work. The better the soil, the more complete the clearing. Dryland salinity soon gave a warning that the tree clearing was excessive, often causing rising water tables that brought salt towards the topsoil. But worse was to come. As in all economic enterprises, attention was focused on the obvious components of the system — in this case crops, pastures, sheep and cattle. Towns were also built, and each of the householders created a garden to his or her individual taste.

In these gardens and on the broad acres new environments were created, more lush than the native woodland with its grasses and shrubs. While these lush pastures provided abundant and nutritious food for sheep and cattle and many pasture-feeding indigenous insects such as beetles, caterpillars and grasshoppers, they did not provide a suitable home for most of the birds, lizards, insects and other animals that are not plant eaters. In the preceding chapter the gradual disappearance of the habitats of these species was described. While these animals can still be found in the remaining fragments of woodland, and their intrinsic value is certainly appreciated, the abundance of many has declined. This reduction is a matter for grave concern, **because the declining species include many that feed on insects which damage pastures, crops and trees**. In some parts of Australia there is no longer a **balance** between the insects that feed on vegetation (the lush pastures, the crops, the garden plants and the remaining eucalypts) and the insect-eating birds, the parasitic wasps, the predacious insects, the spiders and the lizards. In a balanced system these natural controls eat vast quantities of insects that would otherwise become pests. Agriculturalists, in their enthusiasm for high production per hectare, are often blind to the destruction of the habitat of these natural controls of insects, which consequently become serious pests of crops and pastures.

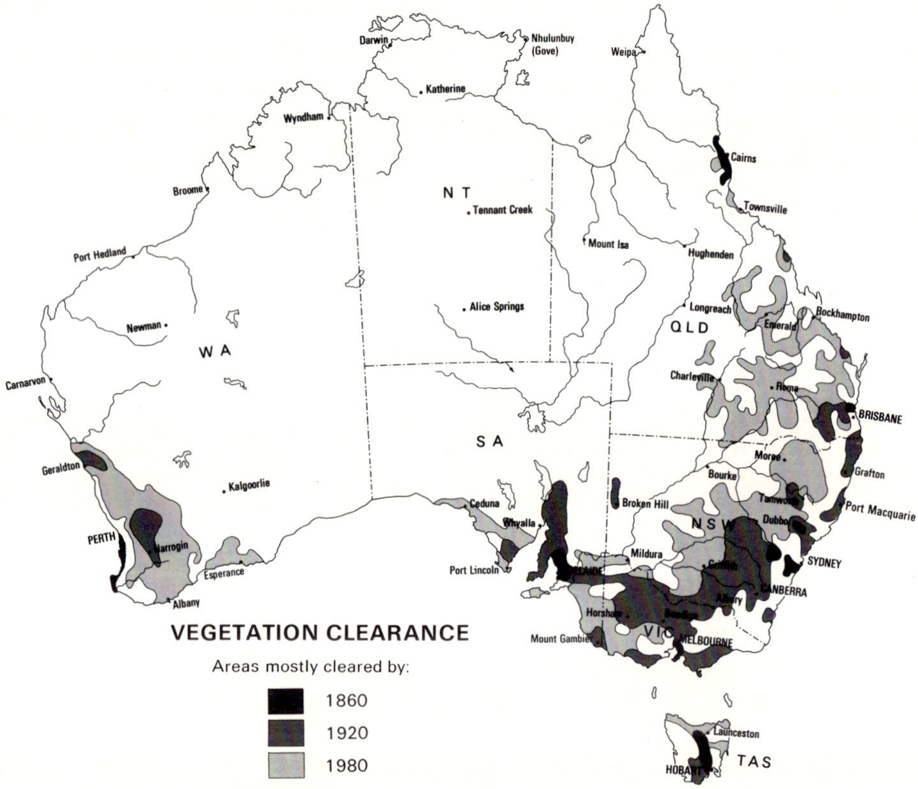

FIGURE 19

Clearing of woodland, shrubland and forest accelerated greatly after 1860, partly due to the shift to wheat growing and the extra labour available after the gold rush. Large areas were ringbarked or cleared with axe and fire and, in the twentieth century, with heavy machinery.

By courtesy of the Australian Surveying and Land Information Group.

On the tablelands it is surprising how many of these useful and fascinating species still persist. However, few of them will be seen in treeless improved pastures, because their habitat is lost and the environment is inhospitable. Many of the native predators and parasites have persisted only because fragments of eucalypt woodland remain. Today it is still unusual to find a 100 ha paddock without a tree in it, but as the old trees disappear treeless pastures will be commonplace. Many paddocks have patches of eucalypt scrub on rocky outcrops, but these remnants are the last remaining footholds for many animals in a fragmented ecosystem.

Let us examine this fragmented ecosystem more closely. Woodland remnants that have an understorey of shrubs and a litter of leaves and fallen logs on the ground support thousands of species of native animals. About 50 species of birds feed on insects in New England. Hugh Ford has recently found that there are between 20 and 30 birds (of about ten species) in an average hectare of undisturbed woodland. Each day these eat 250–500 g of insects from one hectare. Depending upon the time of year, the lizards (mainly skinks and dragons) number roughly 200 per hectare, more when there is more litter. These lizards would eat about a thousand insects daily from this hectare, including their eggs and larvae, spiders, worms and so on. Wasps, robber flies, dragonflies, ground beetles and wireworms would catch and eat or parasitise many thousands of insects each day in the same hectare. Spiders together would catch or snare perhaps a thousand more insects daily; species such as flies, beetles, moths and grasshoppers. When we add these numbers together, it becomes clear that in the woodland ecosystem a balance exists between the insects feeding on plants and the insect-eating animals. Although this balance fluctuates through the seasons and between years, there is an ecological stability in the long term. The diversity of animals and plants somehow contributes to the stability of communities such as eucalypt woodland.

During the first 100 years or so of settlement an increasing percentage of woodland was cleared. It was as if the warp of the woven fabric of the eucalypt woodland ecosystem (the trees and shrubs) had been largely removed, leaving the weft (the native grassland). Although a little threadbare, the fabric held together. The sheep and cattle, however, were beginning to damage the structure of the remaining woodland. In particular, the palatable shrubs like wattles and dogwood were being eaten by livestock. Furthermore, burning was periodically destroying the long grasses, litter and tree trunks that were home to so many lizards, spiders and small birds that nest near the ground. Then, from 1950 onwards, the threads of the weft were cut out, metaphorically speaking, and the holes in the weave were filled with patches of introduced evergreen grass and clover. As stock numbers increased from 1960 onwards in step with the higher productivity of pastures, the remnant woodland was trampled and grazed more heavily by the stock. **If there was a fault in the management of tablelands pastures, it was simply that on most properties the woodland remnants were not fenced off from livestock.**

We all know what has now happened. The threads from which the ecosystem pattern was woven have unravelled as one species after another has dwindled. There is very little comparative information on animal populations in grassland, woodland and partially cleared woodland. Hugh Ford and Andrew Bell have counted birds in such contrasted habitats around Armidale, and the results are striking. In Eastwood State Forest there were 21 birds per hectare in August–September; during the same period there were ten birds per hectare in open woodland on the New England University campus; while in a dieback area at Newholme there were only two birds per hectare and treeless grassland had less than one bird per hectare. The number of bird species recorded in each of these habitats was respectively 46, 35, eight and six. We wish we could have found information about other classes of animals in the tablelands, because it seems that the fragments of the original ecosystem are falling apart. Dieback of the eucalypts is just one of perhaps a hundred changes that have occurred without being recorded, because they are less noticeable.

Unfortunately research over the years has become so specialised that very few people have looked at the agricultural ecosystem as a whole. Nor do the disparate pieces of research add up to even the beginning of an understanding of the ecosystem in which animal production is practised. We are not being critical of the individual excellence of the research workers, but critical only of their separateness, each concentrating on his or her own field. Recognising the need for a holistic approach to research and development, the Australian Government is providing funds for 50 cooperative research centres (CRCs). Thirty-four of these CRCs have been established where federal, State, university and private enterprise research will be integrated (*CRC*

Compendium 1992). Seven of these CRCs will investigate agricultural topics, namely tropical pest management, biological control of vertebrate pest populations, tropical plant pathology, plant science, legumes in Mediterranean agriculture, soil and land management and catchment hydrology.

Research on pastures and animal production has been going on for 70 years, yet so little research has been done on the effect that grazing and trampling are having upon the tablelands ecosystem. John Begg published an article in 1959 which showed that over five years the soil under improved pasture was drier than that under native pasture. In 1973 Jim Langlands and Ian Bennett made a passing reference to the fact that water infiltration rate declined with increasing stocking rates in New England (fig. 20); the rate of water infiltration in pasture grazed at 17 sheep per hectare was only a third of the rate at 2.5 sheep per hectare. Surely there should be a comprehensive study of the relationships between stocking intensity and water run-off and erosion?

It is pointless to apportion blame, but in the face of the impending disaster of continuing tree decline and soil deterioration, it is essential to be completely honest. If there are possible consequences of tree clearing, pasture improvement and heavy stocking **which were not anticipated**, let us all now examine these possibilities openly. This frank appraisal is by no means easy; it is bedevilled by conflicting interests. No graziers would want to feel that they may be responsible for damage to the environment resulting from their management practices going back 50, 100, or 120 years. Some will vehemently resist an examination of these practices. Yet it is in their best interests to have research done on the effects of livestock on run-off, eucalypt regeneration, pasture persistence and weed encroachment (see fig. 20). In 1986, a review of the human impact on Australian soils was given in a book edited by John Russell and Ray Isbell; and in 1991, William Lines gave a history of the conquest of nature in *The taming of the great south land*.

One of the achievements of the New England Dieback Research Appeal was the awareness in scientists that they were contributing to the understanding of a complex problem of importance in our everyday lives. There was great satisfaction in solving together a problem of great economic and aesthetic magnitude, which had not been tackled elsewhere because the

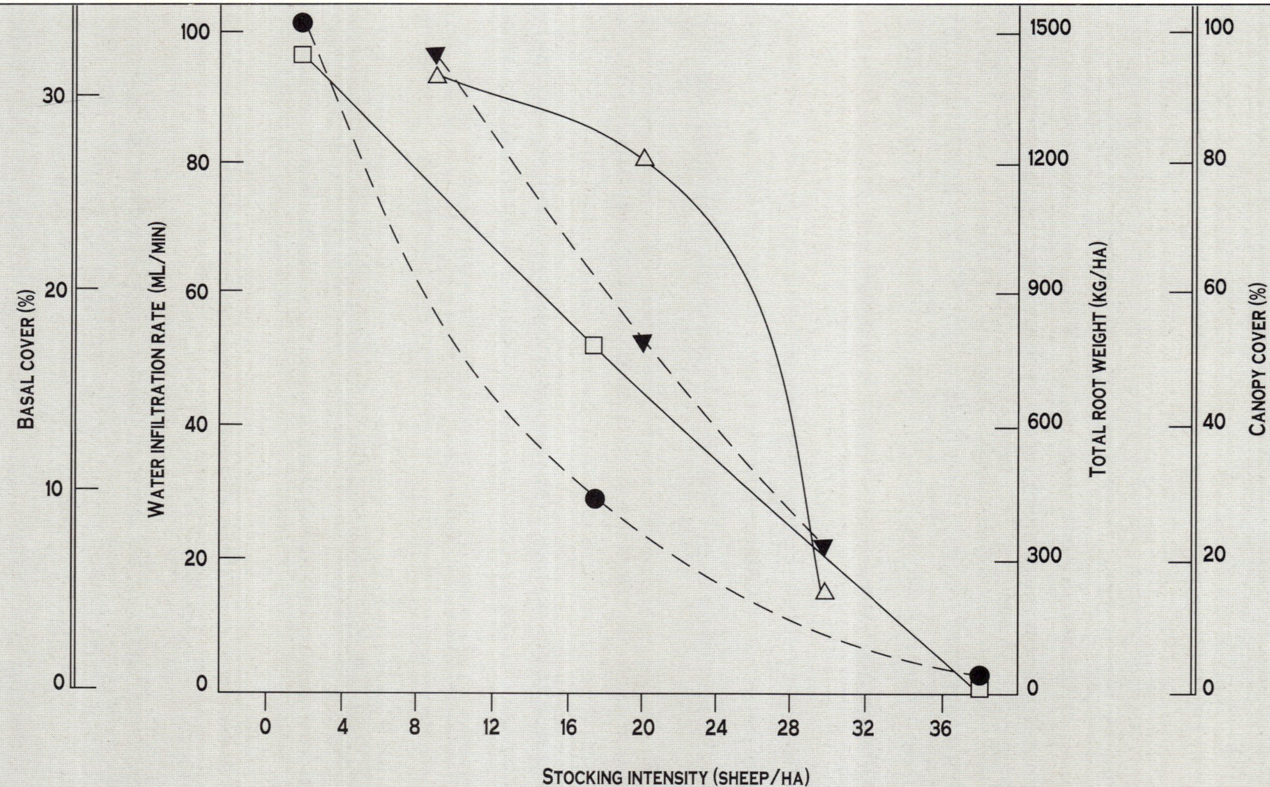

Figure 20

Water infiltration rate (●), total canopy cover (□), basal cover of sown species (△) and total root weight (▼) at various stocking intensities with Merino sheep. Data for the period 1963–68, except basal cover and roots after treatments maintained 1964–73. From Willoughby and Davidson (1979), by courtesy of Cambridge University Press.

PLATE 32

Bearded dragons put on a frightening show if interfered with but it is all bluff — albeit effective.

PLATE 33

Jacky lizards are usually seen sunbaking on logs or stumps where their camouflage affords protection from predators. From such vantage points they, in turn, can keep an eye open for prey on which to pounce.

PLATE 34
Sugar gliders feed on insects, nectar and sap in trees and shrubs. They require hollow branches for nesting. Photo by Roger Fryer.

PLATE 35
Scorpions are Jekyll and Hyde characters. Shy and secretive by day, at night they prey mercilessly upon other invertebrates such as insects and spiders.

PLATE 36
Frogs prefer moist habitats, where they catch such things as mosquitoes, flies and beetles with a long tongue attached to the front of the mouth to extend its reach.

PLATE 37
Chrysomelid beetles display a seemingly endless variety of colour and pattern.

PLATE 38

The first attempts at clearing timber resulted in a massive regrowth of suckers and seedlings of eucalypts.

PLATE 39

In this picture there is a story of nearly 100 years of pastoral endeavour. The stump in the foreground and others like it are remnants of handfelling in the earliest days. Later the ringbarking of trees of all sizes was found to be effective in preventing sucker growth. In this scene the living trees are the third generation after two complete clearings.

PLATE 40

On many grazing properties eucalypt woodland remnants are grazed and trampled so heavily that regeneration of trees is suppressed. The fate of these old trees is inevitable.

PLATE 41

Christmas beetles are incorrectly blamed for most of the damage to eucalypts. The larvae occur in soil under grassland and the beetles fly to eucalypts in late December through to early February, where they feed on the edges of leaves.

PLATE 42

A mountain white gum with an understorey of rough-barked apples and wattles. The litter from the peeling bark and falling leaves provides a habitat for many ground-dwelling insects, because the soil is cool and moist. Burning in years past destroyed this habitat.

PLATE 43

Blackthorn flowers in summer and autumn providing nectar for myriads of useful wasps and flies. Because this colourful bush is not browsed by cattle or sheep, it can be planted in grazed paddocks. Many small birds nest in blackthorn bushes.

PLATE 44

Rough-barked apple trees are covered with a profusion of flowers in midsummer.

prospect of research on such a broad problem had daunted individual scientists. When the appeal was launched, the promise was made that the research would be closely integrated and directed towards a practical solution to the problem, to be used by landholders who wished to reverse the decline of trees on farms. The solution may apply elsewhere, because dieback occurs in many parts of Australia — in Western Australia, in Victoria, throughout the New South Wales tablelands, in southern Queensland, in South Australia and in Tasmania. There has also been an uncontrollable dying of trees in overseas countries, but it is less spectacular than dieback on the northern tablelands. People did not stir themselves to say: 'We must stop this change we are causing'.

Trees do not die without cause. No doubt the reasons are complex, with many contributing causes. But with the whole of New England as the laboratory, and the enthusiasm displayed for the research, the means were found to reverse the changes. Eucalypts in their natural setting should be established wherever the landholder chooses. Some graziers may choose to have no trees at all in their paddocks. There is an apocryphal story about a prominent grazier who was asked as he felled the last gum tree in a lush pasture: 'Don't you think the cows would appreciate the shade?'. He replied: 'They shouldn't be lying in the bloody shade, when they could be out filling their bellies with grass'. But most graziers elected to have shade trees and windbreaks and even some extensive areas of woodland left untouched during the ringbarking years. The aim of all the research on dieback was simply to give people who own the land the same choice over again — a second chance. No-one is going to tell them what to do, but only how to do what they choose.

This choice has to be made by the people who own the land — the graziers and the farmers. Individually they are obliged to make a farm management decision, and then bear the cost and face the consequences. A policy of laissez-faire is one of the available choices. Frankly it is the attitude we have all taken until now. Let us try to visualise the way we are headed. In a matter of decades there will be extensive areas of pasture without trees, because the tall shade trees will have died if current signs are any indication. The fragments of woodland will be grazed so that no seedlings can survive, and **one after another the old trees will die without being replaced.** The woodland will thin itself until the scattered trees will be insufficient to give safe breeding havens for even the adaptable magpies and magpie-larks. In later years we will rarely hear or see the wildlife described in this book — unless the decline of trees can be reversed.

As a direct consequence of the loss of the natural controls of insects, there will be increasing damage to pastures, crops and also trees by the innumerable insects that are indigenous to Australian farmlands. There are over 40 species of scarabs, whose grubs feed in the soil under Australian pastures. There are several species of underground grass caterpillars (species of *Oncopera*) that are present over wide areas in all States. There are many species of root weevils, cutworms and wireworms as well. There are crickets and grasshoppers of many species. These numerous indigenous potential pests are occasionally locally abundant, and sporadically they cause severe localised damage to pastures. But the general level of abundance is usually kept below seriously damaging levels by the wide range of insect-eating animals described in these pages, and the climatic stress described in chapter 4. This dynamic balance will be lost if the remaining patches of eucalypts are allowed to disappear. Graham Goodyer, in an Agfact pamphlet on pasture scarabs on the northern tablelands, describes the predators, parasites and diseases that contribute to the present favourable balance. Graziers are advised to fence off native vegetation on farms to exclude livestock and allow the trees and shrubs to regenerate to aid pest control.

A parallel can be drawn with New Zealand pastures. There, the early settlers deliberately destroyed the natural vegetation, leaving hardly any vestiges of natural forest or scrub on farmland. Peter Pottinger explains that New Zealand's improved grasslands are disturbed systems and consequently are not conducive to the establishment, build-up and maintenance of natural enemy populations. Attempts have been made to introduce predators, including the Australian magpie and wasp parasites from Australia. Starlings have been encouraged with nesting boxes. Now that the inexpensive organochlorine insecticides like DDT and BHC can no longer be used, because they lead to toxic residues in milk and meat, the cost of treating pastures with modern insecticides to control grassgrub (a scarab) has risen to between $48 and $200 per hectare for each application, excluding labour and machinery costs. These organophosphate pesticides are only active for a year, and treated pasture cannot be grazed for several weeks after chemical application. In addition, resistance to each chemical will develop because the kill after each application is less than 100 per cent. So it can be seen that the use of insecticides against soil insects is expensive and often ineffective. In Australia the cost would be no less, and such controls are obviously out of the question. But this is the road along which we are travelling. Are we to repeat the mistakes of the New Zealand pastoralists? We are presently drifting in that direction.

The habitat of the wildlife that is quietly going about its task in the woodland and surrounding pastures is being destroyed. In the late seventies, tree dieback seemed to be proceeding uncontrollably, having been set in train by causes that we did not then properly understand. **The perception that the dieback problem is irreversible must be carefully analysed because it will determine the future of trees on farms, and the course of farming history.**

Trees in woodland die from many causes, such as age, waterlogging of soil from rising water tables or impeded drainage, salinity, insect damage from borers or leaf-eating insects, disease and, importantly, damage from livestock. Less frequent causes are frost, drought, lightning, rabbits, hares and so on. Isolated trees and shrubs are more likely to suffer frost damage and smothering by weed growth, especially young trees. Livestock are the main threat to trees on farms.

Let us now consider the insects which are found on eucalypts. Trees are damaged by indigenous insects of many species. Chrysomelids (tortoise or leaf beetles) of several kinds may be found beneath loose bark on eucalypts in winter. In spring they emerge to feed on eucalypt leaves and later lay their eggs like a bracelet on a fine twig of a tree. The larvae feed on eucalypts, and the mature beetles lay the eggs of the second generation which do further damage in the autumn. Sawflies, those wriggling clumps of sticky larvae that exude a distasteful yellow fluid when disturbed, are another pest of eucalypts. At night they disperse to feed on the leaves. Leaf skeletonisers are the larvae of an inconspicuous moth; the caterpillars feed on the tissue between the leaf veins. The cup moth larva, with its paired clusters of hairs, feeds on eucalypt leaves; the pupa is encased in a tiny pear-shaped cup attached to a twig. Several other caterpillar species feed on eucalypt leaves — some tie the leaf-tips together, others roll the leaves and one forms a silken case with which it moves around. Some are tiny and feed between the upper and lower epidermis of the leaf; others are as long as your hand.

Table 2

Seasonal Prevalence of Insects on Trees as a Source of Food for Predators and Parasites

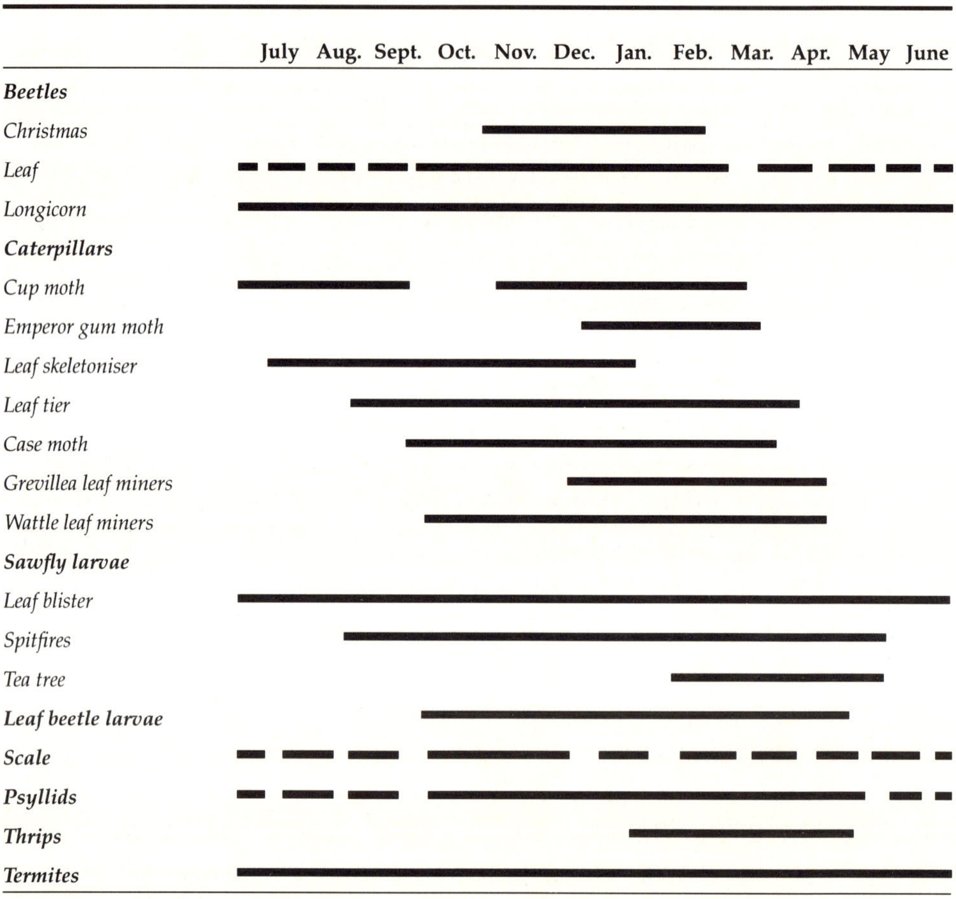

Source: Hadlington and Johnston (1988), reproduced by courtesy of New South Wales University Press.

Phasmids or stick insects feed voraciously on leaves; although solitary, they occasionally do serious damage. Sap-sucking insects, although small, may kill the young growth on eucalypts and deform seedlings by injuring the growing tips. Naked scale insects sometimes cover the young branches; armoured scale insects form hard scales or galls on the leaves and twigs. Some coccids form large woody galls on branches of saplings. The sugary manna found under eucalypts is the lerp of a sap-sucking psyllid; lace lerps will be found on the foliage of eucalypts, together with the little insects that form them.

All these insects live entirely within the eucalypt trees, and bear no relationship to the pastures. Collectively they form the food supply of the insectivorous birds found in woodland. But these insects also feed on isolated trees, and particularly on seedlings of eucalypts in pastures. Here the birds are far less common, and no doubt this is one reason why the scattered eucalypts are so badly damaged. As well as the birds in woodland, innumerable insect parasites and predators occur and these, together with many diseases, have established the ecological balance which ensures stability in eucalypt woodland. These insects are an essential part of the ecosystem, the various species providing food for birds throughout the year. Remember that birds do not hibernate, so they need food every day of the year.

In table 2, the seasonal prevalence of the common insects on trees and shrubs is shown, the data taken from Phil Hadlington and Judy Johnston. Unless a wide variety of trees and shrubs is present on a property, birds will not be attracted to find food and breeding places.

In some years, when insects are scarce, insect-eating birds do not breed. Many nectar-feeding birds also require large quantities of insects and spiders to feed their young. In all countries it is known that birds breed at those times of year when insects are abundant. For this reason we find it strange that some scientists are seeking eucalypts that are unattractive to insects. We should not be concerned that trees do not grow as fast as they might without insects. After all, the trees are not a crop to sell, but a habitat providing additional food supplies to biological controls of farm pests. When birds are excluded from branches of eucalypts by enclosing the branches in nets, the number of insects is greatly increased (see fig. 21).

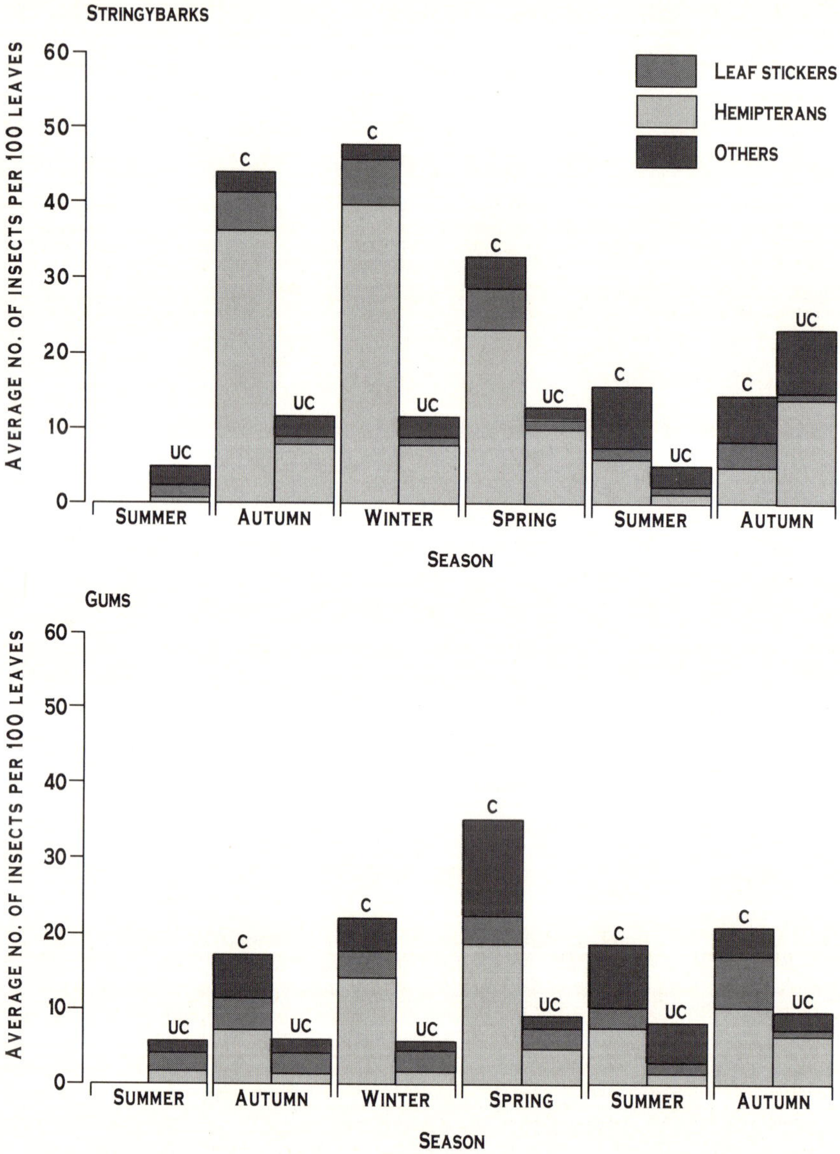

FIGURE 21

Numbers of insects counted on branches from which birds were excluded (C) and those to which they had access (UC) in two types of eucalypts (Dunkerley unpub.).
From Ford (1986), by courtesy of Surrey Beatty & Sons

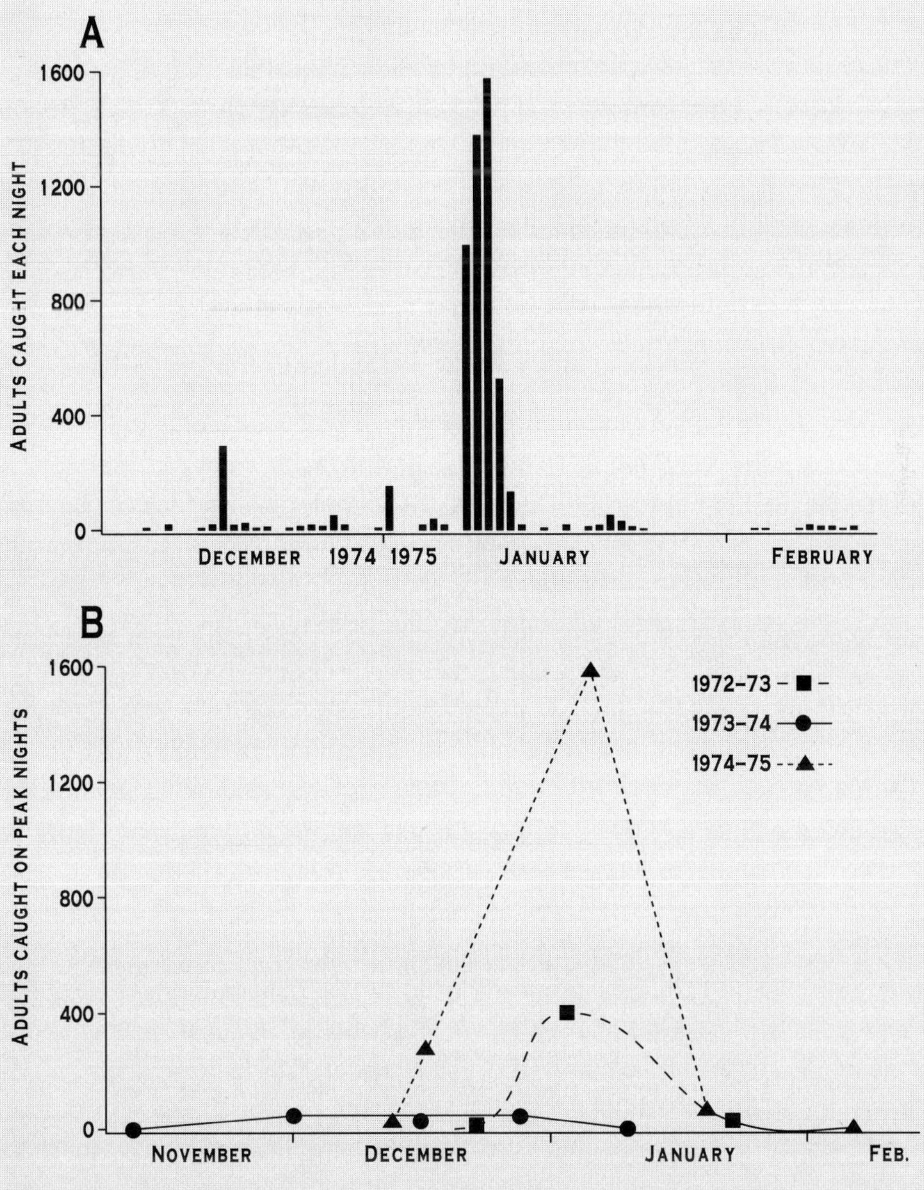

FIGURE 22

Nightly light trap catches of Christmas beetles in a red gum tree at Armidale (A), and the duration of high activity in each season over three years (B).

The scarab beetles that feed on eucalypts come from white grubs that feed and complete their development in grassland soil. These grubs ingest soil and roots and digest the bacteria and fungi on the organic matter, eructing methane gas much as ruminants do. Christmas beetles (*Anoplognathus* species) fly to eucalypts and remain there day and night, feeding and copulating. Because they are so colourful they are noticed by all and sundry, and blamed for dieback out of all proportion to their limited role. Fortunately, Christmas beetles feed for only a few months each year, in late December to February. Peak catches in light traps occur on only a few nights of the year (see fig. 22). The other scarabs that damage trees are the smaller brown beetles (*Sericesthis* species) that fly only at night, and the damage these do is limited by their becoming inactive on cool or windy evenings, when they return to the soil. If cold, wet weather occurs during the peak flights in December and January, and the beetles cannot feed, the females resorb their eggs and lay only a few eggs that season. The large brown scarab beetles that come to lights in summer do not feed at all; these are species of *Rhopaea* and *Antitrogus* which have rudimentary mouth parts and neither males nor females can feed.

According to light trap data, Christmas beetle activity has declined markedly since light trap records were first taken by Reg Roberts at Chiswick near Armidale, and later by Gus Campbell at nearby Arding. The total seasonal catches of *Anoplognathus* and *Sericesthis* from 1963–64 to 1990–91 are given in figure 23. Phil Carne, from a study of *Anoplognathus montanus* on the southern tablelands, expressed reservations about light trap data for estimating beetle abundance, but concluded (on the basis of tree defoliation trends) that Christmas beetle numbers had neither increased nor decreased between 1966 and 1976. It is important to be clear on this point, because it has been repeatedly suggested (wrongly, we think) that Christmas beetles are one of the main causes of dieback. One of us has spent 20 years researching scarabs, including a study of fecundity and mortality of the species. Although scarab grubs grow larger in improved pastures, and the beetles produce more eggs, this potential increase is counterbalanced by a higher mortality in drier, hotter soil under treeless improved pastures, as described in chapter 4.

The quite negligible damage (with a few exceptions) inflicted in the past decade on eucalypts by insects of all kinds has shown up the difficulty of using insecticides as a preventive measure; that is, treatment in expectation of an

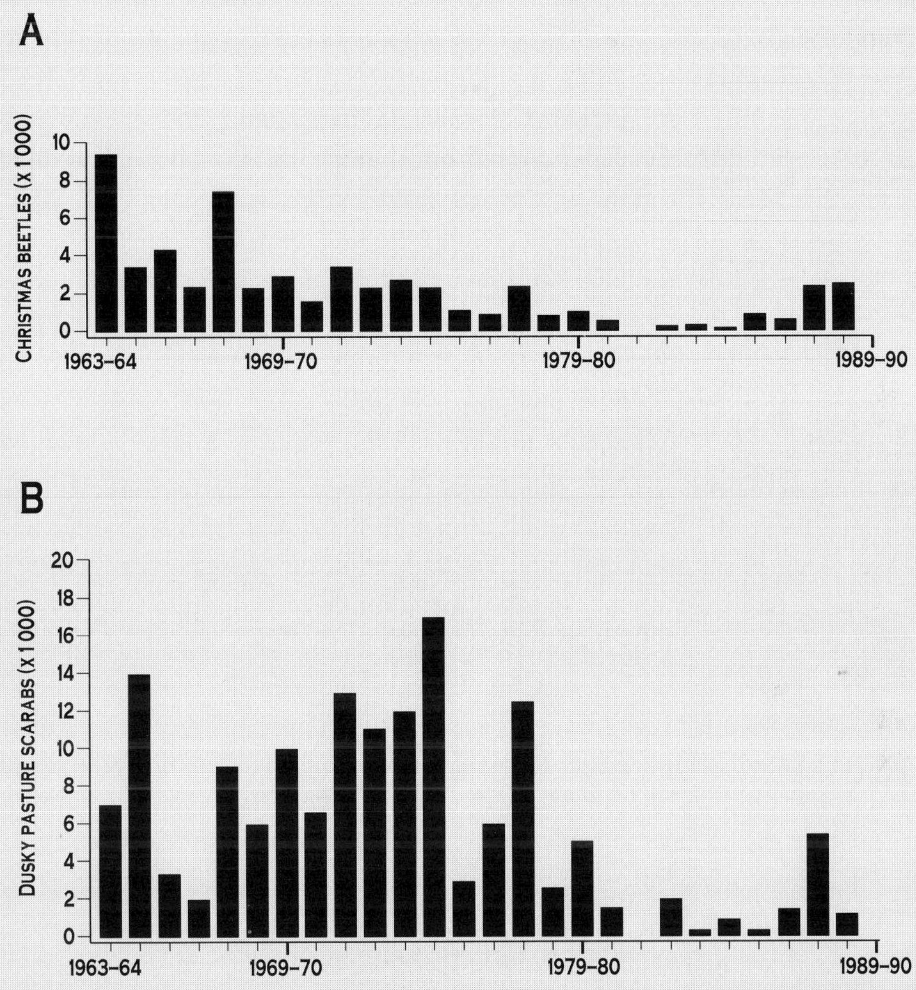

FIGURE 23

Total season catches of Christmas beetles (A) and dusky pasture scarabs (B) 1963–64 to 1988–89 near Armidale. Data from Reg Roberts and Gus Campbell.

outbreak which may not occur. One suggestion for preventing damage to eucalypts by leaf-eating insects is to drill holes in the sapwood and inject a systemic insecticide, which is then translocated to the leaves. Protection is given for a month or so. In seasons of low insect numbers all this effort and expense is wasted. Few people are in a position to forecast insect numbers in advance, and such over use of toxic material is deprecated. Widespread use of insecticides for the control of insects on eucalypts is inadvisable, because of the likely adverse effects on the wildlife that is associated with eucalypts. Far better to rely on predators and parasites to reduce pests to acceptable levels. Robin Barker and Wilhelmus Vestjens report that at least a dozen bird species include Christmas beetles in their diet, and more than 100 bird species eat scarab beetles of all kinds.

The research done at the University of New England, with financial support from the New England Eucalypt Dieback Research Fund, has been described in the book *Dieback: death of an Australian landscape* by Hal Heatwole and Meg Lowman. The depressing title hides a thorough review of dieback, but we do not interpret the data as the authors did in 1986. Further research has proved that the recovery of eucalypts after defoliation (often associated with dieback) is vigorous and complete in most cases, so that the tree is not deformed or set back in the long term. Meg Lowman artificially defoliated eucalypt saplings more than five times, and they continued to produce foliage. Young and old trees that suffer leaf fall during drought often shoot vigorously later; or if the trunks of young trees die, new shoots grow from the base of the trunks. The lignotuber (a swelling between stem and root in many eucalypt species) assists in the recovery of these trees.

In simple terms, our view is that the decline of trees on farms (rural dieback) is an imbalance between the rate of tree regeneration and the rate of tree mortality. Death of trees will always occur, caused from time to time by the factors already mentioned. Many of the trees left standing from the earliest years are 100–150 years old. Some survived the extremes of rainfall shown in the 1858–1991 rainfall graph (fig. 18). Some succumbed to drought or flooding, for example in the 1964 and 1982 dry spells and the exceptional wet years 1949 and 1950. Localised outbreaks of insects that defoliate trees have been recorded in New England since 1857, as mentioned in a review by Robert Boyd of mortality of native trees in woodland ecosystems.

There is no evidence that the severity of tree-defoliating insect outbreaks has increased. We stress again that outbreaks are localised, and if you are looking for insect damage you will usually find it somewhere. But records taken from any one of these sites have failed to demonstrate an increase in any tree-feeding species of insect over the past 50 years. Jill Landsberg, Jock Morse and Partap Khanna have compared insect numbers in lightly grazed remnant woodland and heavily grazed pasture with comparable tree density. They found that insects defoliating Blakely's red gum were more abundant in the latter, posibly because manure from livestock under the trees enriched the leaves of the eucalypts. Most of the trees defoliated recover completely, but a small proportion will die. Fungi will infect a few trees but, with the exception of jarrah dieback (caused by *Phytophthora cinnamomi*) in Western Australia, fungi are not a major cause of tree death. A very useful perspective of rural tree decline is provided in Jill Landsberg and Ross Wylie's summary of the stresses causing tree death. Most of these are undesirable side effects of an intensification of pasture management.

To get a dynamic ecological perspective of trees on farms all the causes of tree mortality should be seen in relation to the rate of tree regeneration. If the rate of mortality exceeds the rate of regeneration, the farmland will slide gradually into tree decline.

This decline began on the northern tablelands in about 1950, when pasture improvement began with the application of superphosphate and seeding with clover and temperate grasses. In figure 24 the stock-carrying capacity is shown in relation to percentage of pasture improved on farms. With less than 20 per cent improved in 1949, only 2.5 sheep equivalents grazed on each hectare and tree seedlings were not all eaten. On fully improved properties in 1966 about 7.5 dry sheep equivalents were carried, and the total livestock numbers on the tablelands had risen to 8 million livestock units. On these heavily stocked (and very productive) pastures there was hardly ever a seedling eucalypt or wattle to be found. Not only did the sheep nibble the tree seedlings and the cattle destroy the saplings regenerating from lignotubers, but the dense grassland smothered the eucalypt seed that would have germinated. By 1976, livestock numbers had risen to 10.5 million sheep equivalents and tree decline was most severe on the most highly improved and heavily stocked properties. Jack Sinden, Alan Jones and Peter Fleming, in an economic survey of 19 properties, found that 43 per cent of the woodland remaining in 1973 was suffering dieback. This increased by one third to 55 per cent in 1980. In 1980, all trees were dead or dying in 37 of the 192 paddocks examined; there were no saplings in 65 per cent of the paddocks; and no shrub understorey in 75 per cent of the wooded paddocks. Surprisingly, an increase in the amount of superphosphate applied did not result in any increase in dieback — in fact the reverse tended to be true!

We cannot escape the conclusion that tree decline is caused by sheep and cattle preventing tree and shrub regeneration. The term 'dieback' is a vague and unsatisfactory word, but in the sense that it is used to denote loss of trees on farms (rural dieback), we would say that rural dieback is caused by sheep and cattle. The balance between vigorous native bushland on farms and dieback is tipped by sheep and cattle, not insects or disease.

Think back to the then terrifying regeneration of trees and shrubs after the ringbarking from 1880 to 1920; all the insect species were there in those days, but all these causes of tree death were negligible in the face of that vigorous growth. Ringbarking sessions in the first 30 years of this century had much the same result, until eventually the livestock got on top of the trees. What has changed over the 150 years has been the suppression of tree regeneration by livestock, not an increase in the causes of tree mortality. Over the past ten years, in paddocks where stock have been excluded, we have seen regeneration of eucalypts, wattles and blackthorn as vigorous as ever recorded in the early days.

During the launching of the appeal for funds for the New England Eucalypt Dieback Research Fund one of us made a personal promise to the contributors: the funds would be used to find a practical solution to the problem, so that those who wanted to establish native woodland on their property would be able to do so. The research fund has served its purpose admirably and the last twenty-odd thousand dollars were used in what came to be called the *Eucalyptus* Regeneration Program. The aim was to demonstrate and trial various methods of native tree and shrub establishment on farms on the northern tablelands. David Curtis and Chris Nadolny were employed on this project. In conjunction with the community-based Armidale Tree Group and a dozen landholders and their families, about 30 000 native trees and shrubs

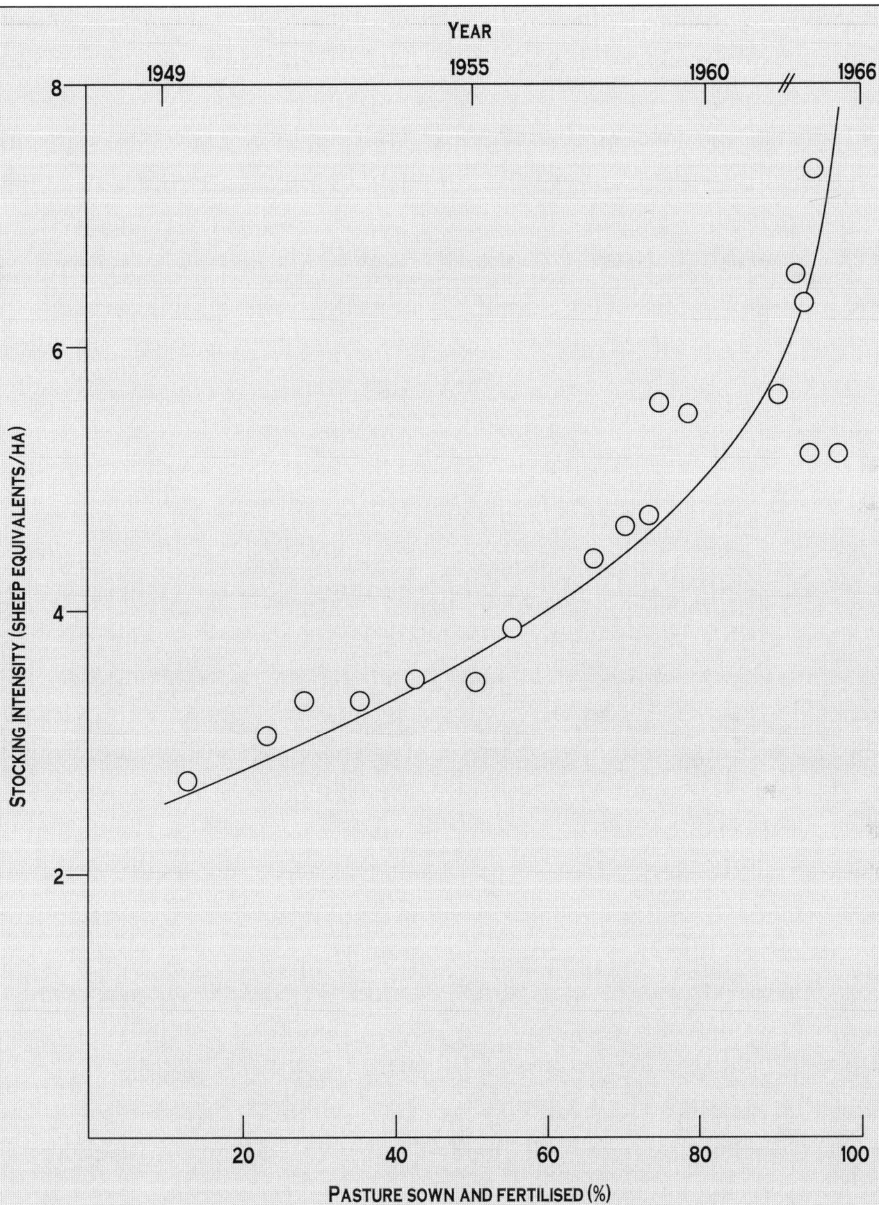

FIGURE 24

Relationship between stock carried (sheep and cattle as dry sheep equivalents) and percentage of pasture sown and fertilised on 11 survey properties between 1949 and 1966, drawn from data in Cook and Malecky (1974). From Willoughby and Davidson (1979), by courtesy of Cambridge University Press.

were planted in dieback-affected country between Guyra and Walcha from 1985 to 1991. In the first years of the trial survival rates were unacceptably low, but as the technique was refined survival improved, to be as good as survival rates in plantings in the more hospitable environments. Later plantings have also been successful, and the growth of many of the eucalypts has been very encouraging.

The *Eucalyptus* Regeneration Program, supported by federal government funds and substantial inputs from the landholders, still continues. More recently, several landcare groups have been established on the tablelands, as well as the regional Northern Tablelands Tree Management Committee, designed to develop a strategy to reverse tree decline. Increasingly, landholders are developing property plans to address tree decline, supported by extension officers from the Department of Conservation and Land Management, New South Wales Agriculture, the Forestry Commission and Greening Australia.

For some reason known only to themselves, some entomologists do not accept that damage to trees by insects is, in most instances, a temporary setback which is quite acceptable and can be alleviated by providing habitat for natural controls. Many foresters have been swayed by those working on insects defoliating trees, and when the National Afforestation Program began the Trees on Farms demonstrations on the northern tablelands, the guidelines were to plant conifers and native species in a 50:50 ratio. The survival of conifers and natives at the second monitoring on the 14 properties is summarised in table 3 (see Suzanne Gaynor 1991).

Some people choose to plant Monterey pine (*Pinus radiata*) and cypresses (*Cupressus* spp.) for windbreaks on farms and we would not question their choice, with one reservation: **if they are choosing exotics because they believe that eucalypts will not survive,** they will inevitably neglect the eucalypt woodland remaining on their property. Throughout Australia, native woodland and bush remnants and scattered eucalypts, rough-barked apple, kurrajong and the like provide almost 100 per cent of the tree cover on farms. If farmers and graziers turn their backs on this native vegetation it will eventually decline from neglect and, once it is lost, the country could never afford to revegetate with exotics the hillsides, streambanks and bleak farmland.

Table 3
Northern Tablelands National Afforestation Program — Trees Planted 1988–90 and Percentage Survival in 1991

Property	Total planted	Natives		Exotics	
			%		%
		Monitored	Survival	Monitored	Survival
Adgin Green	*11 297*	*4 684*	*89*	*4 403*	*93*
Berranee	*6 200*	*2 359*	*84*	*3 602*	*91*
Blaxland	*4 552*	*1 545*	*87*	*1 362*	*57*
Chiswick	*13 165*	*4 208*	*79*	*6 843*	*62*
Cressbrook	*4 839*	*1 543*	*98*	*2 769*	*92*
Dunvegan	*18 682*	*1 129*	*59*	*15 792*	*76*
Glen Hope	*1 835*	*827*	*85*	*404*	*79*
Linwood	*7 204*	*506*	*57*	*6 296*	*87*
Lyndhurst	*5 109*	*1 969*	*89*	*2 414*	*84*
Mickey Gully	*2 795*	*511*	*84*	*1 734*	*77*
Newholme	*9 779*	*4 916*	*90*	*551*	*70*
Sandon	*2 809*	*486*	*87*	*1 439*	*80*
Swallowfield	*4 182*	*3 196*	*86*	*1011*	*80*
Woodlands	*3 670*	*1 942*	*84*	*1 272*	*55*
Totals	*96 118*	*29 821*		*49 892*	
Means			*85*		*78*

Remember that, worldwide, every continent has in parts allowed domestic livestock to destroy native vegetation to the point of barrenness: the Middle East, Africa, Asia, Europe (including Britain), North and South America. In New Zealand, the early settlers deliberately used the hoof and tooth of sheep to clear regenerating woodland. The saying was: 'Let a sheep carry an acre, until an acre can carry a sheep'. The Queen Elizabeth the Second National Trust was established by New Zealand's parliament in 1977 to facilitate the retention and management of native vegetation on farms, and the response was enthusiastic. We hope to persuade landholders that native vegetation has a unique value on farms in Australia.

There are many large grazing properties on the northern tablelands where 5–10 per cent of the land was retained under woodland after the final round of clearing in about 1920. On many of these properties the woodland was not fenced to exclude grazing, and today there is not a single seedling of eucalypt or wattle. Let's take a look at a typical eucalypt woodlot. On 'Saumarez' station near Armidale there is a stretch of woodland of more than 200 ha, and from 1976 to 1981 Sue Bowen studied in intricate detail the germination of plants there. The woodland had young saplings beneath the canopy of older stringybarks, gums and rough-barked apples, showing that eucalypt woodlots can be successfully maintained on farms. However, a detailed measurement of more than 1000 trees showed a disquieting fact. Not a single seedling of *Eucalyptus* or *Angophora* was found in five years. Figure 25 shows the percentage of trees in each stem-diameter class. The smallest trees (less than 1 cm diameter) are shoots from root stocks many years old. We must conclude, therefore, that although up to about 1970 there was abundant regeneration (witness the fine saplings 1–10 cm in diameter throughout the woodland) there had been no germination of eucalypt seed after 1970. Sheep have always been allowed to graze lightly in the woodland, and only recently have cattle been excluded. By contrast, across the road where a sparse woodland had been spared from grazing for a few years, the grassland was interspersed with seedlings and suckers of eucalypts, rough-barked apples and wattles. On parts of 'Saumarez', too, exclusion of livestock from small areas in recent years has enabled New England peppermint and other eucalypts to grow into sizeable trees.

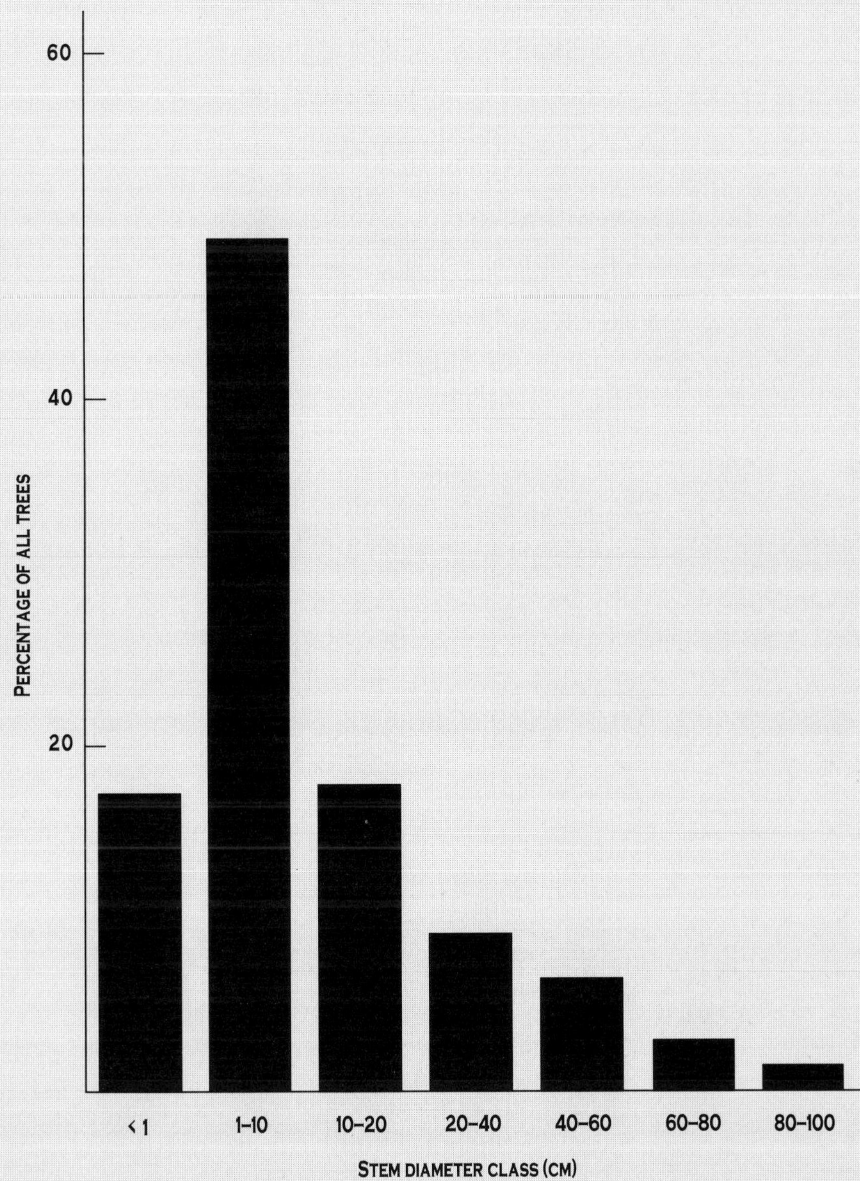

FIGURE 25

Trees in a farm woodland remnant grazed by sheep, showing absence of seedling eucalypts; the stems less than 1 cm in diameter were shoots on lignotubers.
Data from Sue Bowen by courtesy of the *Armidale Express*.

In every farming region of Australia a few far-sighted landholders have excluded livestock from remnants of native bushland on their properties. Some of these are described in chapter 14. Without exception, these remnants are thriving woodland ecosystems — an inspiration to us all.

We hold a firm view on the nature of the countryside that appeals to us. Such views are quite personal, and we have exposed our feelings to encourage others to make their choice. Such choices are very hard to make, because every person will bring a special mixture of sentiment, economics and prejudice to the judgement. However, the choice lies between perfectly clear-cut alternatives — you either accept the value of native bushland in your farming or reject its unique value.

Bushland is not incompatible with modern intensive grazing management, because a good stock-proof fence will counterbalance, in time, all causes of rural tree death by encouraging tree and shrub regeneration. So finally we have answered our own question: 'Yes, each one in turn does have a choice'.

13

Exercising Your Choice

Our aim has been to find out all we could about the habitats of the predators and parasites which are natural controls of the many insect pests (some very serious) in pastures, gardens and eucalypt woodland. A second objective developed when it became apparent that there was an obligation to answer the question in the title of the book. This is a personal opinion, our carefully considered scientific judgement. Our answer is simply: 'Yes, you have the choice of retaining a powerful array of biological controls, because you can save the eucalypt woodland as a habitat on your own property — thriving fragments of woodland between your improved pastures and crops'.

At first, we did not think that an unequivocal answer could be given. However, we were surprised and delighted to find so much life in the fragmented ecosystem. Until then we had not counted the ibises feeding day after day in pastures. Nor had we seen 50 scoliid wasps on one blackthorn bush, or a dozen spiders in a small clump of grasses.

For some years now we have watched very closely the dying of eucalypts, but at the same time we have noted the seedlings of stringybark, red gum, yellow box and New England peppermint springing up in hundreds beneath the dead trees. So when others were speaking despondently of dieback, we tempered our concern with the realisation that regeneration of eucalypt woodland is not impossible.

Many will be surprised that our answer is unequivocal. However, the exercising of this choice requires careful planning, a few changes in farm practice, and above all a determination that is sustained over many years. Unless this effort is made soon, the task will become more difficult. The remaining fragments of the ecosystem on which regeneration must be based will be further reduced. The time will come when the task of regenerating the woodland habitat as a source of natural controls of farm pests will be insurmountable.

In the previous chapter the choices open to landowners were spelled out. You now have to exercise your choice. This book was not intended to be your conscience. Of course we feel strongly about the loss of plants and animals, and we shall do all we can on our own land to preserve them for our enjoyment. But if a neighbour wishes to shoot his crows and ravens, and fell all the timber on his land, we will not question his rights. We only hope that owners of land are aware of the consequences of their actions.

Increasing use of computer models (decision-support systems) to integrate masses of knowledge and help landowners with management decisions is an encouraging development. Peter Gillard, John Williams and Richard Monypenny built a decision-support model to evaluate options for tree clearing and pasture improvement in the semi-arid tropics of Queensland. The tax law that regards tree clearing as a capital expense affects the economic decision in favour of retaining trees and oversowing the woodland with stylos (introduced pasture legumes) and applying superphosphate. But most importantly the model shows that tree clearing carries with it environmental risks to the landscape. These risks can be slow to develop, but when fully expressed they make any short-term economic gain from tree clearing insignificant. This is an extreme example of the most difficult decision facing you: How much of the farm should be covered in bushland? A professional farm planner can help you think this through, but the decision will be yours in the end.

In preparing a farm plan, consider the land-use capability of the different classes of land on your property. This will identify the areas that should remain wooded or be planted to trees because they are water recharge areas or too steep and shallow to be grazed or cultivated. Don't be biased by the present use — remember that you are correcting past mistakes. Drainage lines may require protection and replanting. When you have decided where to cultivate or graze, you should consider the need for shelter and access roadways. Bear in mind the need for corridors to link the bushland that may be scattered across the farm because of the topography. When you contemplate the fencing needed to control livestock remember that you are planning for the future, and fencing can be done over many years.

When you have decided where you want trees, you have to choose between exotic and native species. The future of trees on farms depends on the opinions of landholders, and these opinions will be influenced by the information available to them. We have followed the extension advice and the articles in the press over the past ten years, and what depressing headlines they have been!

'Ecological balance of New England lost forever', *Armidale Express* 1980
'Dieback — the silent killer of the majestic gum: Devastation of our national tree proceeds unheeded, unchecked', *Weekend Australian* 1983
'Beetles turn killers', *Australasian Post* 1989
'Regreening goes awry as insects target eucalypts', *Canberra Times* 1989
'Political trees may never bear fruit, say scientists', *Sydney Morning Herald* 1989
'New threat faces billion trees plan', *The Land* 1990
'Beware, the dreaded lerp!', *Albury-Wodonga Border Morning Mail* 1990
'A faction that plays God with landscapes', *Canberra Times* 1991

But then some of the articles were more hopeful:
'A 30-year commitment to tree planting', *Armidale Express* 1983
'Farmers co-operate on landcare', *National Farmer* 1989
'Tree planting program is an outstanding success', *Armidale Express* 1989
'Farmers battle to save our soil', *Australian* 1989
'10 years on... and the fight against dieback continues', *Armidale Express* 1989
'Our trees: can we put them back?', *Good Weekend, Sydney Morning Herald* 1989
'Rural dieback: restoring a balance', *Ecos* 1990
'Reversing rural tree decline', *Rural Research* 1990

No wonder people were confused!

Some entomologists exaggerate the damaging effect of insects on trees and, as a result, many graziers are now afraid that eucalypts will not survive, and believe that planting them is not worthwhile. Until just a year ago the trees-on-farms extension officers were still uncertain of the wisdom of recommending the planting of native trees.

We began planting trees from local seed in 1980, and after the 1982 drought (which most of them survived) they grew unchecked. They are now large sturdy trees. But equally important was the abundance of self-sown seedlings of eucalypts — stringybark, yellow box, Blakely's red gum and apple box. It was as if the gods looked down, saying: 'For 150 years we thought these

Australians hated trees; now they are planting the same, so let's give them a hand', and they brought the seed and the rain together.

On farms you should be aware that you are not planting just trees, but establishing ecosystems to improve the sustainability of your farming. We have given examples of natural pest control without insecticides — an enormous saving in costs. Over 25 years, this simple method of pest control has proved its worth (see chapter 3). So go out and examine your native vegetation.

- Is it adequately fenced against livestock?
- Does it have an understorey of shrubs and herbs that will provide nectar for control agents in all seasons of the year? Table 4 gives the flowering period of some common species; those that flower in summer are especially valuable.
- Can you find seedlings and saplings of eucalypts, wattles and rough-barked apples beneath the larger trees? If the smallest tree is 10 cm (or more) in diameter, your fence is probably not holding.
- Remember the dictum: If there are no shrubs in the bush remnants on your farm, you are heading for trouble; if there are no tree seedlings or saplings you are already in trouble.
- Leave the dead trees as nesting places for birds, and also the fallen logs that are home to many predators and indeed insects to feed predators when pests are scarce.

When you plan new windbreaks on a property, try to start them near a bush remnant, so that the birds and the predators and parasites of farm pests can work their way along the windbreaks. As you work around the property, keep an eye open for the birds in the trees. You will soon come to recognise them from their calls too.

There are some real difficulties in exercising your choice. One is the dependence of some species on a habitat far larger than even the most extensive grazing property. Ibises, for example, range over 50 km in one day. The poisoning of insects on which they feed, for example by spraying wingless grasshoppers in pastures, could be a hazard. The use of insecticide on potato fields could adversely affect ibises, magpies, magpie-larks and a long list of wasps and other predators.

Table 4
Flowering Periods of Native Trees, Shrubs and Herbs that are a Source of Nectar and Pollen

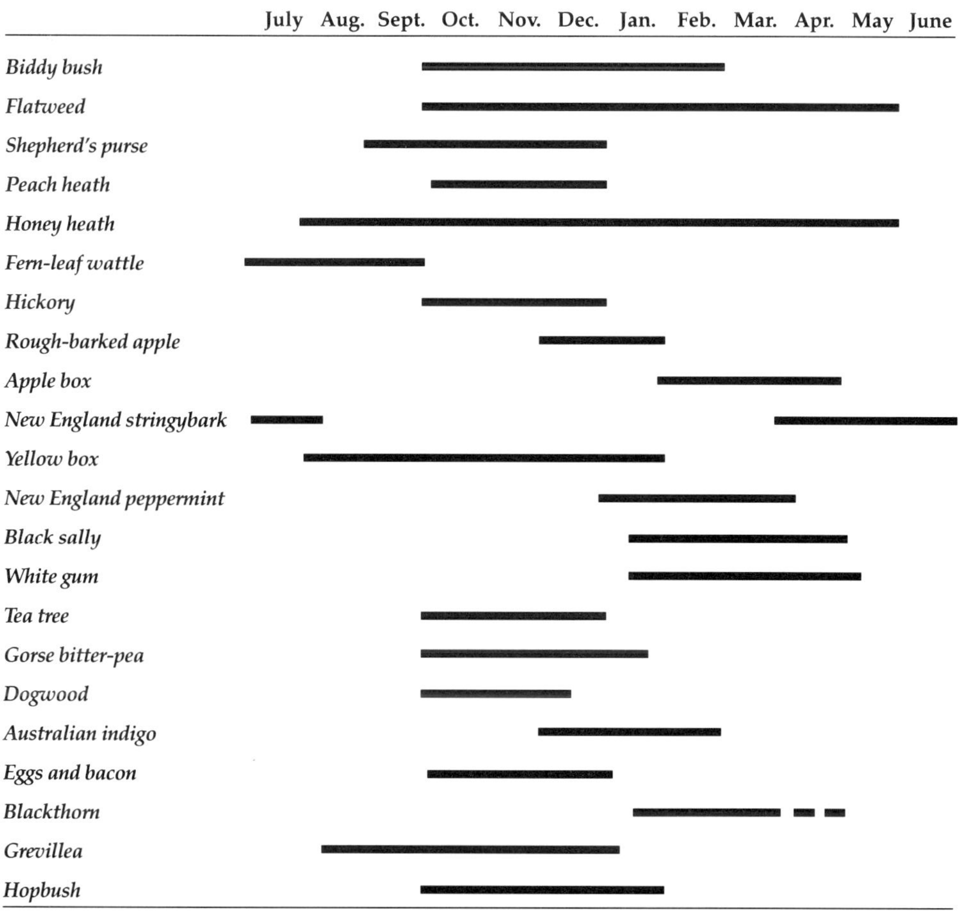

Source: Sue Bowen, pers. comm. in Davidson (1982), by courtesy of CSIRO Australia.

More drastic than these local actions would be the loss of the breeding places of ibises, such as the draining of the Macquarie Marshes in western New South Wales. The management plan for the Macquarie Marshes, prepared by the Water Resources Commission and the National Parks and Wildlife Service in 1985, provides for the allocation of 50 000 megalitres for wildlife habitat without detriment to the other users of irrigation water. For example, the controlled release of water from the Burrendong Dam prevented the drying of the marshes during the breeding seasons for ibises in 1989 and 1990. Building of block dams on the Booligal Swamp has also improved the breeding sites of ibises, to the benefit of farmers for hundreds of miles around.

In a suburban garden you are dependent to some extent upon the trees of the surrounding countryside, which are the breeding places of magpies, black-faced cuckoo shrikes, rosellas and other birds. But they are easily attracted to gardens. Many birds will choose a suitable garden for a home. Some of the lizards will make a small garden their home, choosing the compost heap or woodpile where there is cover. Spiders will use any foothold, yet there are households where a spider is never seen — beware any spider which betrays its presence with a web!

Another difficulty when you exercise your choice should be recognised in advance. The response to new practices will be slow, because the ecosystem needs time for the interdependent populations of plants and animals to reach a new balance. If an area of grassland with scattered eucalypts showing signs of dieback is fenced off from cattle and sheep, there should be a scattering of eucalypt seedlings 10 cm high within a year, provided there is viable seed in the ground from a tree that has set fruit. There might also be eucalypt saplings 1 m high from rootstocks previously grazed and trampled. *Acacia* seedlings will grow to a height of 50 cm where seed is available. In the second year, the dry grass foliage from the previous year will form a surface mulch which will keep the soil moist. Small animals, including insects and earthworms, will increase in number. In step with these populations, lizards, spiders and predacious insects will increase too.

However, there may be unexpected problems. Under shady trees where cattle and sheep have been camping for many years, the accumulation of manure may cause a proliferation of fungi which on some soils create water resistance. Even after the heaviest rain these non-wetting soils will be too dry for eucalypt seeds to germinate. Application of detergents or scarifying the surface soil will solve this little problem.

As the eucalypt seedlings grow into saplings, many will be partly defoliated by the leaf-eating insects described in chapter 12. Although some seedlings will be stripped by caterpillars, beetles or sawflies, most will survive these attacks. But as the predators and parasites increase in the new and favourable habitat created by your management decision to exclude domestic stock, most of the eucalypt saplings will grow unscathed. Healthy eucalypts, the stems straight and the leaves hanging clean on the pendulous young branches, are a sight you may have forgotten. The seedlings that have died or are damaged beyond recovery by insects are of no concern. You need only a few hundred eucalypts per hectare, and in places you may have many hundreds of young trees. Do not try to save them all, but see them and the insects on them as a source of support for predators that can protect other eucalypts and crops and pastures from insect pest attack. That is, beneficial predators and parasites cannot survive without a source of food — such as leaf-eating insects — and these, in turn, may kill some trees. This doesn't matter as long as the ecosystem as a whole functions well.

If you can be convinced of this from your own experience, you will have acquired an understanding of ecology that eludes many professional agriculturalists. You will see a new balance between pests and biological controls being slowly re-established, being eased in the direction you want by your management decision. Remember that most predators and parasites must feed every day, and that in addition to living trees and shrubs, the litter composed of dead grass, twigs and bark on the ground provides habitat for many small animals that serve as a continuous food supply for predators and parasites, especially during droughts.

There will be good years and bad years for trees to grow. Ironically, the droughts which depress the growth of improved pastures are favourable for the growth of many native plants, unless the drought is severe and prolonged. In the hot dry weather some insect pests are adversely affected; for this reason the growth of eucalypts is exceptionally good. But there will be bad years in which eucalypts seem to be the target of a dozen insect pests. The only consolation is that damage will be less in the fenced enclosures than the damage to eucalypts growing in isolation in grazed pastures.

Research which has been done on eucalypt dieback and bush regeneration has answered many of the questions that will arise when landowners begin their own program of regeneration of woodland. If necessary, you can find out about nutrient requirements of trees or about competition with grassland. This book has described insect pests and their natural controls. You will know from experience and observation in which areas it is easy to grow trees, and on the other hand where regeneration is difficult to achieve. In some places there will be no alternative to planting seedlings or reseeding. In this case there is the difficulty of selecting species, but the wisest choice is to put back the species that grew naturally on each site. Where no remnants can be found, our advice is to form a landcare group, and ask a field botanist from a local university or government department to prepare a feasibility plan.

John Williams has published lists of species associated with typical New England soils, slopes and aspects. On each geological formation there is a catena (or pattern) of species in the region, as one goes from the top of the hillside to the valley bottom, depending on depth of soil, drainage, subsoil, rock outcrops and fertility. Local provenances (or varieties) of trees have developed on these sites, and locally-collected seed usually produces vigorous plants when planted on similar sites. Computer programs, such as the SELECT program developed by John Ive and Doug Cocks, can supply this information. It is surprising how widespread natural regeneration is, including on the notorious New England tablelands, provided stock are excluded from the selected sites.

When it is clearly understood that the aim is not to grow a few trees for shelter, but to establish a complex ecosystem of hundreds of plant and animal species, the need to choose sites close to existing woodland will be appreciated. It is fortunate that most farms have not been stripped of all woodland. We have expressed our pleasure in the wealth of species in these fragments of woodland. You in turn have expressed alarm that these fragments are slowly disappearing. This book tells all we could find out about the fragmented ecosystem and its value to good farming and pleasant living. This chapter was added to strike a note of optimism, and to ask you to use the information now at hand to make a start.

Whether or not Australian farmlands regain their character and natural beauty depends on landholders. Who knows, perhaps small communities throughout Australia will achieve what no other country in the world has achieved.

14

Success Stories Australia-Wide

We began writing this book ten years ago, seeking to understand the forces that would determine the fate of bushland on farms. Most of all we have looked at things from the viewpoint of farmers and graziers, fully aware that the future of the countryside will be decided by them. Much has changed since then.

During this time there has been a growing concern among governments, farmers and townspeople about the decline of trees on farms. There has been a surge of tree-planting activity by landholders, often in Landcare groups assisted by extension officers. Townspeople have volunteered to help to plant trees and prepare seedlings, for example the Armidale Tree Group. The Australian Trust for Conservation Volunteers is a well-organised group of this kind. There are also individual farmers who have been quietly going about the task of planting native trees and shrubs on their land for 30 years and more. The National Tree Care Award in 1985 attracted 260 entries, and some of these are described briefly in *Trees: why we need them countrywide,* edited by Nina Riemer.

Case studies in environmental hope, edited by Peter Newman, Simon Neville and Louise Duxbury, describes vividly what people on the land in Western Australia are doing to retain native vegetation on farms. The 1988 ABC Tree Care Award provided a wealth of personal stories about people who have improved the habitat for wildlife in Australia and New Zealand. *Grow your own wildlife: how to improve your local environment*, by Peter Johnston and Alan Don, gives thumbnail sketches of these thoughtful people.

This chapter was contributed by people living in all States of Australia, writing about inspired farmers and their elation when they have regenerated native bushland. When they saw the first birds feeding in trees they had planted they were deeply moved. In time, they will feel the satisfaction of knowing that they have planted not just trees, but an ecosystem.

A Concerned Tasmanian

By Elizabeth Bond

Gerald Maxwell Burbury, Max to all who knew him, was born in 1903 in the Midlands area of Tasmania. He was a man of vision, energy and fun who applied the 'work hard, play hard' philosophy throughout his life.

When he first saw 'Ratharney', the property at Woodbury where he lived for most of his life, the hillsides were covered with native trees — she-oaks, wattles and gums. In 1924, however, a huge fire began in the Tiers and burnt through to the east coast, a distance of more than 50 km, before it was stopped. Most of the native trees covering the hills were killed, leaving a bare landscape.

The Midlands area of Tasmania has been showing significant tree decline for many years, the rate increasing markedly in the last 10–15 years. Owners of one property near 'Ratharney' estimate that 90 per cent of their trees have died in the last ten years. While all the factors causing this tree decline are not known, it is felt that lack of moisture after ten years of below average rainfall has put severe stress on trees and shrubs. The stressed trees are then not able to overcome the effects of browsing animals and insects, such as the large increase in numbers of brush-tailed possums and, in some cases, sawflies, and so they die. This is happening to trees and shrubs of all ages, from newly-planted to old trees, but the high incidence of death in young plants up to 15 years of age is particularly worrying. A local community group, the Midlands Tree Committee, has obtained funding from the National Soil Conservation Program to begin an investigation into some of these factors.

Max Burbury was a man before his time, who recognised the early signs of the problems we now face and decided to do something about it.

The Woodbury area is one of the driest in Tasmania, with a high frost incidence, even out of season, and prone to drought. These difficulties breed tough people, and also make it tough to establish trees.

Along with his contemporaries, Max spent a large part of his young life ringbarking and clearing paddocks. When interviewed by Amabel Fulton, a local journalist, he said: 'When I was young they used to say the only good tree was a dead one. I never liked killing trees although I would when it was necessary, when it was the sensible thing to do to make good paddock land. But I would have left far more trees than the early people did — trees are essential for shade and shelter if you are running stock.'. This outlines an essential part of his approach for at least 50 years.

After returning from World War Two in 1944 he took on the 'Ratharney' property; later adding 'Ballochmyle', a nearby property with similar problems to 'Ratharney'; and 'Racecourse', a higher-altitude bush property with more rain. He suspected that 'clearing trees had buggered up the rainfall', so he began his planting project. From the forties until his death in 1989 he never stopped.

This was a time when the government was encouraging land clearing with tax incentives. There were no 'recipes' for tree re-establishment. Max had a vision and he began experimenting with different ways of getting trees to grow. He tried to create a park-like landscape with single tree plantings, but these needed a lot of maintenance, were prone to insect attack and frequently died.

Groups of trees proved to be more successful, so he tried woodlots, shelter belts and clumps in various shapes and sizes. He found no one configuration was better than any of the rest, so examples of all types can be found.

Max was always experimenting and planning. He planned months ahead what he would plant and where he would plant it. Each year he chose a different area, concentrating his efforts for ease of management, particularly in the first year of growth. Planting took place from May to September each year. Holes were always dug at least one month before planting. The trees were watered in, mulched heavily and guarded. Intermittent waterings were provided during the first summer and autumn.

He had no preconceived ideas about species or provenances. Everything and anything was tried. He thought she-oaks, blackwoods and English trees would be the trees of the future. Hawthorns grow well in this area, providing good shelterbelts and avenues along the road. Pines, elms and golden willows proved successful as well as many natives. He tried to grow golden willows from cuttings, but this didn't work so it was back to seed.

A cunning businessman, he bartered with friends and acquaintances for trees. Friends grew trees or collected seed of local and exotic species for him. Acorns were collected at 'Lovely Banks', about 50 km south of 'Ratharney', and the oak trees now are growing well. He preferred trees which had been grown in the ground, rather than pots, as he found these survived better. He knew all the trees around the property — swamp gums, blue gums, white gums and elms — and took great pleasure in showing them to anyone who was interested.

He undertook a planting of 2000 trees every year, whatever the state of the rural economy. However, in his later years he felt it ought to be more like 10 000 a year to be sufficient for the future. The majority of his plantings are Australian natives, but not all local provenances. The main criterion was that they must be tolerant of frost (especially to out-of-season frosts) and drought.

One of his unusual plantings is a grove of olive trees. He came to know the trees in the Middle East during the war and decided they had qualities suitable for the Tasmanian Midlands. These trees bear bitter black olives but the turkeys and stock appreciate them before humans get the chance.

Most trees were planted to provide stock with shade and shelter. Providing habitats for native birds and insects was a secondary consideration. The main predator of birds in this area is the feral cat. While Max did not specifically try to attract or to keep birds in the area, they are present in significant numbers, despite the cats. It will be interesting to see in a few years time whether the pastures of 'Ratharney' have survived the recent heavy onslaught of pasture grubs (particularly the red-headed pasture cockchafer) better than surrounding properties. While other pasture grubs have been active in the past, this cockchafer has only been in the area for the past five years, with infestations of up to 1500 or so per square metre in the first year. Crows in particular love this grub and there are large numbers of well-fed crows digging up pasture throughout the Midlands.

All special occasions were marked by planting trees. When each of his sons was born, an oak tree was planted in the bull paddock. These have flourished. However, all trees in or near cattle areas needed extra protection because cattle are hard on young trees.

Around one of the dams native trees were planted to attract insects. These in turn would blow onto the water, attracting trout to the surface to provide good fishing while watching the wildlife.

One end of the property has old salt lagoons. The early white settlers used to come here to collect salt for sale in Hobart Town. A salt-affected area has been planted with trees, giving mixed results. Some have died where their roots have penetrated the salty water table. Trees planted on higher ground have survived.

A man of foresight, Max was determined his ideas would be put into action, no matter what happened to him. In the early sixties he had a scare — pains in his thorax. Convinced that this was a heart problem and that he was near the end, he took his workman (still an active member of the staff) around the property and told him what trees were to be planted and where. Then he went to the doctor. The doctor diagnosed gastric ulcers, so with that news, the consummate whisky drinker ordered his celebratory drink 'whisky au lait'.

Over the years many people visited the properties, learning his techniques and working with him to try out other techniques. He made many friends and won many awards. Two of the later awards were in 1985 — AFDI Tasmanian Tree Farmer Award and Special Merit Award for Amenity — and in 1989 the inaugural Bi-annual Tasmanian Greening Australia Tree Care Award, the 'Individual' Excellent Tree Care Project.

Despite all his work, the older gum trees on the property have died in the last few years due to age, accelerated by the factors causing the general decline of trees in the area. Fortunately, younger trees replace them. Max died in December 1989, leaving his two sons, Alec and Ian, to continue the work he began. They have benefited from the lessons he taught them and also have the memories of his enthusiasm for life and all it contains, enthusiasm he has passed on. As the pressure on the environment increases, and plants and animals are subjected to more stress, Alec and Ian have one great advantage over their neighbours — a 45-year start in tree re-establishment. May we all be as kind to our successors as Max was to his.

PHOTOGRAPH 10
The late Max Burbury on 'Ratharney', July 1986.

PLATE 45

Trees, shrubs and herbs flowering at different times of the year provide nectar for birds and insects, which together help to control pests in pastures, crops and trees. (Clockwise, from top left) peach heath, dogwood, tea tree, grevillea hybrid, sticky daisy bush, blackthorn.

PLATE 46

Aerial photograph of corridors of trees planted between the Dundas and Black Range forests in Victoria.

PLATE 47

'Lanark' viewed from north-west in 1963.

PLATE 48

The same view of 'Lanark' in 1989, with the natural wetlands re-established and the Fentons' dream of trees and birds a reality.

Landscape for a City in the Country

By Judy Frankenberg

In 1973 the Australian Government, in association with the New South Wales and Victorian Governments, created the Albury–Wodonga Growth Centre. This exercise in decentralisation on a grand scale was designed to accelerate growth in the area to cater for a population of 300 000 people by the year 2000. The Albury–Wodonga Development Corporation was established to plan and implement the project. The staff of the corporation included town planners, engineers, social planners and landscape architects. The intention was that the new city would provide an attractive lifestyle alternative to the increasingly crowded capital cities, and take full advantage of the rural setting to create a city in the country.

Several of the early environmental studies carried out in the area emphasised the need to reafforest some of the surrounding hills. This recommendation was extended by the landscape architects and developed in an early landscape study into a forward tree planting program that was adopted by the AWDC as part of its policy. Susan Campbell, landscape architect with the corporation, was given the task of implementing the program.

Tree planting therefore became a major part of the early planning design. Forward tree planting required that future development areas should be planted with a blanket of trees well in advance of any actual construction of roads, housing, industrial estates or other urban development. When development commenced in planted areas, many trees would have to be removed, but those which remained would be well grown and greatly enhance the landscape of the new area. The surrounding hills were also to be planted with trees to provide a woodland backdrop for the city and to restore some over-cleared and eroding hill slopes.

The rural landscape around Albury–Wodonga, on which the new urban areas were to be built, was mostly grazing land which had been cleared up to

100 years ago. In some areas, very few trees were left and these were estimated to be dying at a rate of 1 per cent a year. The cause of death was probably a combination of the usual causes of tree dieback all over the country; such as old age, insect attack, ringbarking by stock and drought. There was some regeneration on the roadsides, particularly in recent years with fewer rabbits and stock movements, but virtually none in the paddocks. Some of the surrounding hills were still tree covered but others were heavily over-cleared and showing some signs of erosion.

Before the forward planting commenced, Susan Campbell commissioned an ecological survey of the development area, which was carried out by Judy Frankenberg. By looking at the remnants of the original vegetation which were left along roadsides she was able to map the natural distribution of the local tree and understorey shrubs. This information was used to plan the choice of species in the plantations.

Planting was not confined entirely to local species. A total of 78 species were planted, including 41 eucalypts, 18 wattles, and species of *Callitris, Callistemon*, *Grevillea*, *Hakea* and *Melaleuca*. However more than half the planting was of species indigenous to the area, while most of the other species were confined to the future urban areas where they would become street or garden trees. The plantings on the hills and in open spaces were predominantly local species. Unfortunately not many shrubs could be planted. Particularly in the early years of the program Susan wanted to obtain a quick-growing tree cover to demonstrate that large numbers of trees could be established successfully. For this reason, a lot of wattles were included which, although short lived, were very fast growing and gave an immediate result to satisfy the doubters, as well as providing shelter for the slower species.

The ecological survey recommended that the planting should include habitat corridors to link the surrounding hills and the central Murray River floodplain through the development area. The corridors have been incorporated in open space areas and on creek and drainage lines and road reserves. These trees will encourage wildlife, particularly birds, into and through the urban areas. The presence of natural or semi-natural indigenous vegetation enables many local birds to continue using an area even after it is covered with houses, particularly now that native species are so widely used in gardens. In fact, the bird population will probably be greater than it was on the treeless farm land which the houses replace. A local bird observers' group carried out a series of bird counts in a variety of habitats at the beginning of the planting program and in the future these can be compared with bird numbers in the developed areas.

The forward tree planting was carried out by local contractors who have subsequently built up a considerable level of skill and experience. This was found to be the most efficient and economical way of planting up to 120 000 trees a year. The planting in the future urban areas was generally at 10 m x 10 m spacings. On the hills, in the corridor, and in future open space areas the trees were planted at closer spacings, usually 3 m x 3 m. The trees are in carefully planned groups or corridors, with the rows running along the contours, either along the ridge lines or down the drainage lines.

This pattern provides both visual and practical benefits: reducing wind speed, controlling erosion, creating wildlife habitat and providing a natural backdrop of trees for urban areas. Planting on land which is at present leased for farming is carried out in cooperation with the tenant farmers to enhance the farming operations. These trees will then enhance the urban area which will eventually be developed on these sites.

When the forward tree planting started, 120 000 trees were to be planted in the first year. It was not possible to buy this number of trees (of the species required) in existing nurseries, so the Caramar Nursery was set up to supply all the trees planted in the forward tree planting program. The nurseries already operating in the Albury–Wodonga area were concerned at the prospect of competition from a corporation nursery, so at first Caramar was restricted to supplying corporation plantings only. The corporation policy on planting extends to providing garden shrubs and trees free of charge for all houses built on corporation land. These are also provided by the nursery and are mostly native species. The birds which move through the urban areas along the corridors will therefore also use the gardens, and help keep the insect populations in check. Caramar now supplies trees to local farmers and to some of the commercial nurseries, and collects quite a lot of its own seed.

The forward tree planting program started in 1977 and has surpassed all expectations in its success. Over one million trees have been planted, including urban plantings, and the survival rates have generally been very good. In fact some thinning has already been carried out in areas where the trees were planted at three-metre intervals, because a higher than expected survival rate meant that the trees were far too dense.

A recently prepared management plan has recommended more thinning, particularly of some non-indigenous species that have proved unsuitable. These will be left off future planting lists. The thinned areas will look more natural as the straight lines are broken up. Many of the short-lived wattles are dying out, which will also improve the landscape effect of the plantings by opening up clearings and reducing the straight line formality which is inevitable in large-scale plantations. Some of these clearings can now be fenced to enable the planting of understorey shrubs, protected from grazing.

The planting program had its critics from the beginning. The greatest concern was the perceived fire risk of having forests close to an urban area. The design and management of the plantations has had to take this very real concern into account. All the plantations have ploughed fire breaks when young, which protect the trees from the outside world as well as the other way round. Once the trees have reached sufficient size the plantations are grazed by sheep. This unfortunately prevents the use of small shrub species in most areas, which may contribute to occasional heavy insect attacks (usually on non-indigenous species). However it has meant that although several fires have travelled through forward tree planting areas, they have not been severe; few trees have been killed and there has been some regeneration of wattles.

There are still some complaints from fire fighters that the density of trees causes problems with access, but during the latest fire, on a very hot windy day, it was apparent that the fire slowed down and moved very slowly through a plantation, without crowning, after roaring across an adjacent grassy paddock at a pace that no fire truck could cope with. So far none of the fires have started in a plantation. As the trees mature and the canopy closes, the undergrowth becomes less dense and the need for grazing will be reduced. There will be no cultivation and it is probable that the herbaceous understorey will gradually revert to a higher proportion of native species, even though some light grazing may still occur.

The establishment of the plantations has not been plain sailing. Rabbits have been the greatest problem, particularly on the hills, and eradication programs have been essential before planting. Almost as threatening have been changes in governments and policy directions and on several occasions Susan has had to fight for the survival of the program. There has also been local criticism of the planting (particularly on leased farmland which will eventually be developed) as using up productive land. However, cooperative planning and the increasing recognition that trees improve the productivity of the adjacent areas has enabled most of the required planting to go ahead at least five years before the expected development of the land.

The impact of the trees on the landscape around Albury–Wodonga is now very marked and even the most trenchant early criticism of the scheme has quietened. Several awards have been made to Susan Campbell and the Albury–Wodonga Development Corporation. These include an award from the Australian Institute of Landscape Architects for landscape restoration, and a National Tree Care Award.

Residents are beginning to use the plantations for recreation. Walking and cycling trails have been constructed through them and they are popular for horse riding. The value of the planting is also beginning to be appreciated on a regional level as the problems of rising water tables and dryland salinity become apparent. Several landcare groups have formed in the region and the Albury–Wodonga tree planting is a constant demonstration to landholders that large scale tree planting can be carried out successfully and at a reasonable cost. One area has been planted specifically for a local landcare group.

Susan Campbell has talked at many conferences and workshops, passing on the experience she has gained and the success achieved in probably the most extensive non-forestry tree planting exercise ever carried out in Australia. The greatest testimony to her success will be that, in 50 years, the fact that the trees were deliberately planted will probably be forgotten and the population of Albury–Wodonga will assume that they have always been there.

PHOTOGRAPH 11

Susan Campbell, who has organised the planting of nearly two million trees and shrubs on urban and rural land for the Albury–Wodonga Development Corporation. She developed a forward tree planting program that is a model for large scale landscape regeneration.

Linking Farmland with the Dundas and Black Range Forests

By Rhonda Lyons

Our project is at Gatum, situated on the Dundas Tablelands in the Western District of Victoria, an area of severe land degradation. The Dundas Tablelands have been harshly milled in the past, and have serious salinity and soil erosion problems. Some of our properties have few native trees, and a dearth of varieties of trees and shrubs needed for ecological stability. The continuing decline of our trees through old age and disease inspired our project; we believe it is vital to our farming environment.

Neil Lawrance, a pioneer in our district and throughout Australia in salinity works, realised the problems of our lands could not be solved by a single farmer. Neil acknowledged the true value of the remnant vegetation, not only on our farms but in both the Dundas Tableland and the Black Range. This acknowledgment by all of us led to the formation of our group.

The group commenced in November 1984 with five adjoining landholder families. Its initial aim was the creation of a wildlife corridor between the two forests, but we are now integrating this as part of an overall land management and protection plan for all our properties. The proposal is to establish a series of interlocking corridors of trees and native vegetation so as, in the long term, to establish a network between the two forests. The project covers an area of 4500 ha of farming land, 40 km of government roads, 20 km of unused government roads and 30 ha of Crown Land. The objective of our group is to link the Dundas and Black Range forests with broad corridors of interlinked native habitat areas. The distance between the two State forests is 14 km (as the crow flies), but the total project works are in excess of 41 km of tree corridors on private land and approximately 40 km on roadsides, unused roads and disused railway lines.

Photograph 12

Neil Lawrance in natural regeneration of river red gum after exclusion of livestock. It is confidently expected that similar self-seeding of newly planted trees and shrubs will occur later.

The tree planting associated with our corridors also complements the measures required to control dryland salinity, soil erosion problems and lack of diversity of trees. We are determined to reverse the decline of existing trees and the neglect of trees on our roadsides. We are determined to protect the trees along the disused railway lines in our area because these provide excellent corridors for wildlife. The dream that spurs us on is a green and pleasant farm landscape.

The landholders feel that they can create an environment that balances the agricultural and native ecosystem. It is hoped that our corridors and our own revegetation programs will provide a more stable and productive farm environment. We believe that salinity will be halted in time and our pastures will be more productive. The corridors will be used by birds to move back onto farmland, where they will feed on insects that damage pastures and trees. Other natural predators and parasites of damaging insects will be attracted by the nectar in flowers of trees and shrubs.

The members of our group have already noticed changes on our farms — increased numbers of birds, greater diversity of species of birds. We have even spotted an echidna out feeding at night. This created great excitement for our children.

Another important change to our environment has been the damage to our trees caused by the sawfly. This insect was damaging our newly planted gums to the extent that whole areas were being defoliated. We used to go out and pick the grub off the trees to save the tree. Now, although the grub is still there, the damage is not so severe. No conclusions from this except, maybe, more gum trees for the grub to feed on. Predators may be having an effect on the numbers of sawflies, or maybe the farmers are prepared to let nature work it out. Also we don't seem to have the time now because of increased tree works each year.

At the end of 1991, after six years working as a group, we have planted approximately 50 000 trees (either as tube stock or by direct seeding), put up 30 km of new fencing and sown 700 ha to deep-rooted perennial pastures.

By determination and hard work we feel we are changing the course of local history, reversing the decline of trees on farms. There are moments of pleasure each of us has looked forward to: the first flowers and seeds on the trees we have planted; the first bird taking nectar or insects from these trees, and in the near future nesting in the corridors; and finally the natural regeneration of the original ecosystem, with a vigour almost forgotten.

None of this would have been possible without the links to the Dundas and Black Range forests created by all members of the group. The members of the group are:

Neil and Sue Lawrance, 'Banool'

James and Susie Waters, 'Redwood'

Ivan and Netta Rees, 'Orana'

Robert and Marilyn Lyons, 'Dundas Park'

Peter and Rhonda Lyons, 'Quamby'

PHOTOGRAPH 13
Trees planted in Salty Gully on the Dundas Tablelands, where dryland salinity is a worsening problem.

PHOTOGRAPH 14

Children from Hamilton School planting trees and shrubs in October 1990. Fifty-nine students planted 2000 trees in one hour.

JOHN AND CICELY FENTON: 'LANARK'

BY PRUE POWELL

John Fenton has been described by many as the 'tree man of the Western District of Victoria'. Our friendship is based on a mutual affinity with trees and the subsequent philosophies, that is: conservation and concern for the earth and its delicate ecosystems. Basically, if there are no trees, there is no life. This view was brought home very forcefully when John inherited his father's property in the mid-fifties and moved into the small house with his young wife, Cicely. The contrast of the leafy metropolis and beautiful coloured gardens of Hamilton, where John was brought up, to the open, windswept, climatically vulnerable 'Lanark' prompted him to begin tree planting almost immediately. A rather stern English-born father-in-law insisting that his daughter could not be expected to put up with such conditions was another major motivation.

John admits to trying to establish cypresses for a couple of years without much success before finding Wail Forestry Nursery near Horsham, and the Natural Resources Conservation League (NRCL) at Springvale. He used their sound basic advice on planting and species and experimented with seeds he collected locally. He became a member of NRCL, a life member of the Society for Growing Australian Plants, a foundation member of the the Australian Conservation Foundation and an early member of Hamilton Field Naturalists. He is currently a trustee of the World Wide Fund for Nature Australia, a member of the Australian Forest Development Institute and a member of the initial farmers' tree group formed in 1980, the Australian Farm Trees Regeneration Committee. He has been a member and office-bearer of the local Rural Fire Brigade and farmers' organisations all his farming life.

Cicely supported John in his tree growing by bearing four potential tree planters and supplying cups of tea in the paddock, enabling him to stay on site and get more work done. (Well, if he was out there planting trees he wasn't inside harassing her to make more tree planters, was he?) While John spent all his spare time and money planting and learning about trees, his neighbours looked on at 'the town boy' with great scepticism. However, when he began putting the water back into the swamps in a horseshoe around the house, they knew he was mad. Fifty-six hectares of grazing land went into wetlands. This might have been the time John's defence line began to harden. It was characterised by the phrase 'Go to buggery'.

He was well before his time, but convinced he was on the right track. The game shooters were delighted and would place breeding boxes on his property to encourage a plentiful supply of ducks. Although a strict 'No Shooting' decree was observed on the property, the shooters realised that the more waterfowl encouraged to breed on the reconstituted wetlands, the more ducks there would be to hunt within the wider district. The water areas on the property now are aesthetically pleasing, an excellent firebreak (the property was used as an example plan of fire protection strategies by the Victorian Country Fire Authority and Rural Fire Service), back-up in drought, and an increasingly intricate ecosystem of flora and fauna.

John's greatest prize in his wetlands is an ibis rookery which established itself six years ago on an area that was drained swampland and bare-grazed for 70 years. The ultimate 'icing on the cake' for the Fentons occurred in 1983. After a previously drained swamp was reconstituted, revegetated and completely fenced for 20 years, three pairs of white ibises nested in the reed beds. These were not disturbed but it is thought that about five young were successfully reared. In 1984, 12 pairs nested; in 1985, 25 pairs nested; in 1986, 100 pairs; and in 1987, 250 pairs. In 1988 and 1989 very few ibises nested owing to dry seasonal conditions.

He does not have to use chemicals for pest control, as the birds using the trees and water as habitat help to restore a natural balance in the insect world.

> Numerous insects feed on the roots and foliage of crops and pastures and may cause substantial damage… water snails carry the parasitic liver fluke that kill many sheep each year… A wide selection of birds exist that are effective predators on these pests. A straw-necked ibis or a white ibis will eat around 200 grams of insects such as grasshoppers and crickets per day.
>
> **(Elizabeth Jacka's RMIT report on birds at 'Lanark', 1989)**

There has been an enormous increase in bird species diversity at 'Lanark' since 1956; from 38 species to 143 in 1986. John says that, of the birds at 'Lanark' now, 33 per cent came back with the trees which have been planted and the last 33 per cent came back when the water was replaced. He has only begun to plant trees specifically for timber in the last ten years. Before that, there was no knowledge of suitable timber species to plant in this area. Rod Bird of the Hamilton Pastoral Research Institute (DARA) has furthered this area of knowledge considerably and, along with Conservation and Environment, they have planted many experimental plots at 'Lanark' with John's blessing. This is a valuable and unique example of science and an on-ground farmer at work solving problems and implementing the solutions.

There are now approximately 35 000 planted trees on 'Lanark'. Capital value of trees on farms is illustrated by two valuations done on 'Lanark', in 1983 and 1988, on 130-hectare portions of land.

**VALUED BY DSM-ESTATES;
REGISTERED VALUER NO. 959**

	No Shelter	Shelter
1983	*$ 790/ha = $103 000*	*$ 900/ha = $117 000*
1988	*$ 1110/ha = $144 300*	*$1970/ha = $256 100*

(*Financial benefits of farm trees*, Victorian Farmers and Graziers Associated and Victorian Garden State Committee)

The farm has operated and still does operate on conservation lines. Insecticides and sprays have not been used for 25 years, while fertiliser application has been nought or extremely light for the past 20 years. Although this has been contrary to accepted local farming practice, the income generated from the home farm has enabled John to buy other properties.

John stocks fewer sheep to the hectare than is generally regarded as a good load. He says actual figures have not been kept and the benefits are hard to list in specific economic terms, but in the long run the value in sustainability and finances is evident in the farm's continued productivity and growth. Fewer sheep require less feed which ties in with his philosophy of none or very small amounts of superphosphate. No superphosphate also seems to cut out many stock troubles, for example ryegrass staggers. There is no need to cut hay to carry the sheep through, which means less labour and fewer machinery costs.

Because there are fewer sheep to compete for the grass, the sheep give excellent wool and meat production. They compact the soil less than larger flocks would, which helps stave off soil erosion. The trees are of enormous benefit to sheep in sheltering them, thereby keeping the stress level of the sheep low so more energy is utilised in wool and meat production. Trees also keep water tables down, so minimising salting problems.

John carried on with his 'conservation farming' through thick and thin because he was convinced there had to be a better and more sustainable way of working with the land. As the quality of the land rises through better management and growth of remnant vegetation and specifically placed blocks and belts of trees, so does the quality of life for the people working in that environment. For example, there is a stark contrast between working in exposed treeless yards with sheep or cattle and working in yards protected and shaded by a plantation of trees and shrubs. I know which I would opt for!

Eight years ago, John began Rural Trees Australia because interest in what he was doing was so great it was interfering with the farm work and Cicely said he had to do something about it. Thus RTA was formed with Roger Young and Hamilton-based naturalist Murray Gunn. They began from an office at 'Lanark'. After two moves, the business was positioned in Hamilton and in early 1990 had four full-time directors, two landscape architects, two graphic artists, a secretary and three outdoor planting staff. In early 1990 John retired from RTA, enabling him to devote his full attention to the revegetation of the six family properties that John, Cicely and their children now operate. The work of RTA is still carried on by Roger Young, establishing about 80 000 indigenous trees and shrubs each year from New South Wales to South Australia.

Through his RTA period John and Cicely still showed a few groups of highly interested people around the property, particularly Senior Lecturer in Landscape Architecture at RMIT, Jim Sinatra and his students. He says he is now more able to devote time to expounding his knowledge of rural 'farm wholeness' in a tutor-extension role with RMIT. John feels this is extremely important and satisfying, as these students are the people who will be helping to recreate an ecological balance in the future. He feels an integrated and balanced way of looking at landscaping must be the basis of any project. This means taking more time to get a feel for the area, study its weaknesses and needs (such as soil structure, wildlife, water tables and exposure) and consider the ways it will be used and how it will be in the long-term future.

The farm and John's experience and knowledge are now valuable resources and examples of the truth that agriculture and nature can and must go hand in hand. The farmer must know the land as a living organism and learn to nurture it. Or, as one of John's favourite authors, Thoreau, says: 'We need the tonic of wildness, we need to witness our own limits transgressed and some life pasturing freely where we never wander'.

PHOTOGRAPH 15

Cicely and John Fenton, who have helped so many people to farm efficiently within beautiful natural ecosystems.

Bird list for John and Cicely Fenton's property 'Lanark', Branxholme

1 Great Crested Grebe
2 Hoarey-headed Grebe Br.
3 Little Grebe Br.
4 Australian Pelican
5 Darter
6 Little Pied Cormorant
7 Black Cormorant
8 Little Black Cormorant
9 White-necked Heron
10 White-faced Heron
11 Cattle Egret
12 Large Egret
13 Little Egret
14 Plumed Egret
15 Nankeen Night-heron
16 Australian Bittern
17 Glossy Ibis
18 White Ibis Br.
19 Straw-necked Ibis
20 Royal Spoonbill
21 Yellow-billed Spoonbill
22 Plumed Whistling Duck
23 Black Swan Br.
24 Freckled Duck
25 Cape Barren Goose
26 Chestnut-breasted Shelduck Br.
27 Black Duck Br.
28 Mallard
29 Grey Teal Br.
30 Chestnut Teal Br.
31 Southern Shoveller Br.
32 Pink-eared Duck
33 White-eyed Duck Br.
34 Australian Wood Duck Br.
35 Blue-billed Duck Br.
36 Musk Duck Br.
37 Australian Black-shouldered Kite
38 Whistling Kite
39 Australian Goshawk
40 Wedge-tailed Eagle
41 Spotted Harrier
42 Swamp Harrier Br.
43 Black Falcon
44 Peregrine Falcon
45 Little Falcon
46 Brown Falcon
47 Nankeen Kestrel
48 Stubble Quail Br.
49 Banded Rail
50 Lewin's Rail
51 Marsh Crake

52 Spotted Crake
53 Spotless Crake
54 Black-tailed Native-hen
55 Dusky Moorhen
56 Swamphen Br.
57 Coot
58 Brolga Br.
59 Masked Plover Br.
60 Banded Plover Br.
61 Red-kneed Dotterel
62 Double-banded Dotterel
63 Red-capped Dotterel
64 Black-fronted Dotterel Br.
65 Pied Stilt Br.
66 Wood Sandpiper
67 Common Sandpiper
68 Greenshank
69 Japanese Snipe
70 Sharp-tailed Sandpiper
71 Red-necked Stint
72 Curlew Sandpiper
73 Silver Gull
74 Whiskered Tern Br.
75 Domestic Pigeon
76 Common Bronzewing
77 Yellow-tailed Black Cockatoo
78 Gang-gang Cockatoo
79 Galah
80 Long-billed Corella
81 Sulphur-crested Cockatoo
82 Musk Lorikeet
83 Purple-crowned Lorikeet
84 Budgerigah
85 Crimson Rosella
86 Eastern Rosella
87 Red-rumped Parrot
88 Blue-winged Parrot
89 Pallid Cuckoo
90 Fantailed Cuckoo
91 Horsfield's Bronze-cuckoo
92 Shining Bronze-cuckoo
93 Boobook Owl
94 Barn Owl
95 Spine-tailed Swift
96 Fork-tailed Swift
97 Azure Kingfisher
98 Rainbow Bee-eater
99 Common Skylark
100 Welcome Swallow Br.
101 Tree Martin
102 Fairy Martin Br.
103 Richard's Pipit Br.
104 Black-faced Cuckoo Shrike Br.
105 Scaly Thrush
106 Blackbird Br.
107 Pink Robin
108 Flame Robin
109 Scarlet Robin

110 Jacky Winter
111 Golden Whistler
112 Grey Shrike Thrush Br.
113 Restless Flycatcher Br.
114 Grey Fantail Br.
115 Willie Wagtail Br.
116 Clamorous Reedwarbler Br.
117 Little Grassbird Br.
118 Golden-headed Cisticola
119 Rufous Songlark
120 Brown Songlark
121 Superb Blue Wren Br.
122 Brown Thornbill Br.
123 Yellow-rumped Thornbill Br.
124 Striated Thornbill
125 Red Wattlebird Br.
126 Yellow-faced Honeyeater
127 Singing Honeyeater
128 White-plumed Honeyeater Br.
129 White-naped Honeyeater
130 New Holland Honeyeater Br.
131 Tawny-crowned Honeyeater
132 White-fronted Chat Br.
133 Striated Pardalote
134 Grey-breasted White-eye
135 European Goldfinch Br.
136 House Sparrow Br.
137 Red-browed Firetail
138 Common Starling Br.
139 Olive-backed Oriole
140 Masked Wood Swallow
141 White-browed Wood Swallow
142 Dusky Wood Swallow
143 Magpie Lark Br.
144 Grey Currawong
145 Australian Magpie Br.
146 Australian Raven
147 Little Raven Br.

Br. = Breeding

The order and common names follow Pizzey's *A field guide to the birds of Australia*.

List compiled by Murray Gunn, Hamilton, 24.2.87.

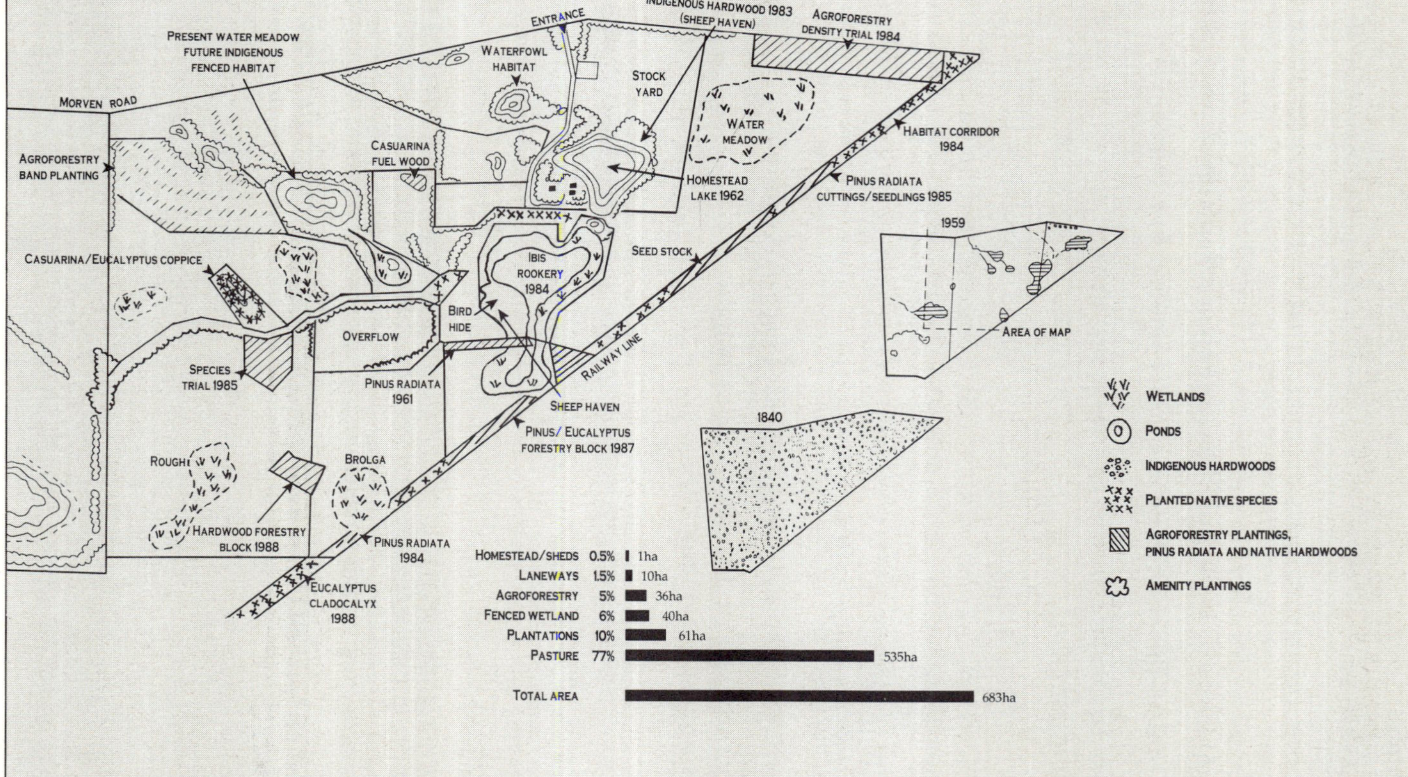

FIGURE 26

Part of John and Cicely Fenton's property 'Lanark' at the time of European settlement in 1840, as it was when John inherited the property in 1959, and as they have developed it today. From Johnston and Don (1990), by courtesy of Greening Australia.

Special People in a Special Place:

How the Williamses of Badgingarra Saved Unique Plants for All Australians

By Jill Reading

Joy and Don Williams are wheat and sheep farmers on the sandplains of Badgingarra, in Western Australia's wheatbelt. They make their living from using the land, but you do not have to be with them long to know that their souls are nourished by their love of native vegetation. They have set aside 323 ha of their 1984 ha property as a nature reserve, preserving seven very rare plant species including four rare and endangered eucalypts, one of which is found nowhere else.

The couple and their two sons tend this huge garden of nature with passion and delight and not a small amount of regret that the exigencies of economics prevent their total immersion in it. 'In a way, it is a pity you have to earn a living — you could do so many wonderful things', comments Don wistfully. By anyone's standards, however, the Williamses have already done many wonderful things.

Both born of sandplain farming families, they began farming the Badgingarra property, 'High Vallee', in 1966. It was a conditional purchase block — land released by the government for agriculture under a number of conditions, including that it be cleared at a fairly rapid rate. Although there had been warnings about the hazards of land degradation on the delicate sandplain soils, both government and farmers were imbued with a gung-ho attitude that meant 'good' farmers cleared their land almost bare to get as much production area as possible. When the Williamses began conservation farming — fencing to soil type, preserving native vegetation, reducing their use of agricultural chemicals to an absolute minimum — other farmers were horrified. Recalled Joy: 'They thought we were mad — the attitude then was clear all, take all, money, money, money'.

Several factors were critical in the development of the Williamses' conservation consciousness. The first paddocks Don cleared and ploughed were deep sand of the kind that is now notorious for developing the land degradation hazard of non-wetting soils, a very difficult problem to reverse. The couple quickly decided never again to clear such soils. On the farm's better soil types, they worked at keeping about half of each paddock in bush to reduce the erosion hazard and to increase protection for stock. 'We try to farm with native vegetation, rather than against it', explained Don. Joy declared her passion for native vegetation by standing in front of the bulldozer when Don was about to clear near the house. Finally, a botanist carrying out a contract survey for the Department of Conservation and Land Management in 1980 showed the couple the many rare species which their land supported, and from that moment they were captivated.

They have since carried out a fencing program costing thousands, with intricate, meandering fences to separate soil types and to exclude stock from plant populations. Every day, if possible, they explore their huge garden, exclaiming at new discoveries, indulging in lively conflict over a correct botanical name, sharing anew their wonder at this special place on earth. Interest from outside snowballed as the Williamses' property — which adjoins Mt Lesueur, known as the State's home of rare plants — became known for its equal uniqueness. So a further sacrifice has involved the time needed to host scientists, students, wildlife societies, friends and visitors of all kinds.

But the sacrifice is willing, explained Joy: 'It does not belong to us — it is really for all Australians'. They were recognised for their endeavours in 1989, when they won the State's premier acknowledgment for preservation of native vegetation, the John Tonkin Tree Awards Gold Award. While, said Don, some farmers around them still think native vegetation is a challenge to be rooted out at all costs, the Williamses are now unofficial advisers to many locals keen to combine farming with conservation. But they do not consider themselves experts. 'The more you know, the more you know you do not know — how to be humble, that is what "High Vallee" has taught us', says Joy.

PHOTOGRAPH 16

Don and Joy Williams, 'High Vallee' nature reserve, with Mount Leseuer mallee, found only there.

Conservation 'On the Land'

By Julianne Venning

Although conservation has become a catch-cry in rural areas in recent years it is not uncommon to find farmers who have been dedicated to conservation most of their lives. One such farmer is John Smyth from Salter Springs in South Australia. He has set aside and managed areas of native vegetation on his property for 25 years and also successfully experimented with a wide range of tree establishment techniques including tree planting, direct seeding and natural regeneration.

His first attempt to re-establish the remnant native vegetation was in 1965, when he fenced off 12 ha of degraded peppermint box and South Australian blue gum woodland, now known as the Aviary, in the hilly non-arable eastern portion of his property. The understorey had degenerated because of regular grazing and only old golden wattles and native boxes were present. He had hoped that once the grazing pressure was removed the native understorey would regenerate as it had in the adjoining Flagstaff Reserve.

In 1962 his parents had excluded all stock from 28 ha of woodland on the eastern perimeter of the 480 ha family property. The area, dominated by peppermint box and blue gums, had previously been used only for summer grazing. Regeneration of the understorey was prolific and spectacular. John believes that exclusion of the stock coincided with a decline in the number of rabbits and this enabled the understorey species to proliferate. The colourful displays of the orchid *Caladenia* enthused John and led him to initiate his own project in 1965. Flagstaff Reserve is a unique area in the lower mid-north of the State and in 1983 was placed under a heritage agreement to protect it in perpetuity.

Although there has been some regeneration in the Aviary it has been nowhere near as spectacular and John attributes this to the greater degree of degradation of the understorey before the stock were excluded.

In the early seventies he established two plantations in the house paddocks. He used mainly the Western Australian species coral gum, swamp mallet, blue mallet and Strickland gum which were very popular at the time. As the plantations began to establish, John noted an influx of birds around the homestead which included honeyeaters, mistletoe-birds, wattlebirds and Adelaide rosellas.

The next phase of John's planting efforts began in 1982 when he was asked to participate in direct seeding and natural regeneration trials run by the South Australian Department of Environment and Planning. With financial assistance from the department, John fenced off direct seeding plots further along the creek from the house paddocks, towards the hilly eastern portion of the property. John's property provided a challenging site on which to establish the local native species by direct seeding.

The site had been used for growing crops for nearly a century and the use of fertilisers and crop species had created a much altered environment. The introduced species are able to establish quickly on the exposed high-nutrient soils following ground disturbance and suppress or even prevent growth of the slower-growing native species. Weed control was essential for successful re-establishment of the native species and therefore the site was ideal to evaluate a wide range of treatments. Native vegetation along nearby roadsides and on the property was a valuable source of seed for the trials.

Out of interest John carried out a variety of treatments at the site which included:

- cultivation;
- cultivation followed by application of a residual herbicide;
- cultivation followed by application of a selective knockdown herbicide;
- grading away the uppermost 2–5 cm of topsoil to remove the weed seedbank; and
- application of a residual herbicide before sowing seed along riplines where the treated soil had been pushed aside.

Seedling establishment was closely correlated with the effectiveness of the treatment. Cultivation, with or without pre- and post-emergent herbicides only contained the weeds for a month or so. Grading greatly improved percentage establishment as the removal of the weed seeds in the topsoil retarded re-invasion of the plots for up to a year. Residual herbicide also improved percentage establishment compared with the cultivation treatments because, like grading, it provided effective weed control during the first year.

While the trials were in progress John further extended the plantations along the creek by both planting and direct seeding. He used the direct seeding technique which at that stage had been most successful in the trials. In May–June 1984 following the break of season he applied a mixture of amitrole and atrazine in one-metre wide strips. About a month later he made a furrow approximately 20 cm deep in the treated strip in which to place the seed. The seed was applied at a rate equivalent to about 4 kg per ha. It was mixed with dry sand to allow even spreading of the finer eucalypt and melaleuca seeds. With this technique the furrow is usually created by using a ripper, but as John did not have a ripper he used his grader blade tilted almost perpendicular to the ground.

By the end of winter there was an abundant germination of seedlings of all the species used in the mix. Eventually John would like to join the remnant vegetation in the hills to his plantations on the flat to encourage the movement of birds across his property to the homestead.

The department also established an exclosure adjacent to the Aviary in order to monitor natural regeneration of the tree species. The fenced area was sufficiently large to allow the seed shed from parent trees to fall within the exclosure. Weed control to create a weed-free seedbed was carried out, using knockdown herbicides, soon after the fence was erected. Ground preparation preceded those periods when seedfall, temperature and rainfall were thought to be favourable for regeneration. During the five years of the trials no seedlings were found though, since then, a few have established.

The lack of seedlings further confirmed John's view that once the native vegetation has become highly degraded natural regeneration is at best a slow process. This is clearly illustrated by the differences between the three exclosures now established on John's property. Further, the lack of regeneration of the tree species (particularly peppermint box) suggests that favourable conditions for their regeneration may occur infrequently and the chances of success are limited in an environment that has been substantially altered.

Early in 1987 members from Trees for Life (TFL), the South Australian Branch of Men of the Trees, approached John about preparing a farm plan for his property as a bicentenary project. John's interest in tree planting was well known and it was suggested that preparation of a farm plan would establish a clearly defined framework for future plantings.

The farm plan aimed to integrate land management, agricultural production and proposed property developments so that sustainable production from the land could be maximised while land degradation would be minimised. It provided an overview of the property to ensure that the right trees were grown in the right place to achieve their intended purpose. In this instance few changes to the farm layout could be justified.

An inspection revealed that the property could be considered in two parts: the rocky hills along the eastern boundary, which cover approximately one-third of the total area; and the flat land of highly productive heavy soils used for cropping. Scattered remnant trees cover the hilly areas and provide excellent shelter during the lambing season. The heavier soils of the flat land are consistent throughout and are not prone to erosion. Roadways allow for ease of access to each paddock. He was happy to initiate further plantings because he likes the look of a well-treed property. However, as there is ample shelter on the property for stock, John was unwilling to fence off substantial areas for windbreaks on the highly productive cropping land.

Five areas were identified for planting to add to the aesthetic appeal of the property and ensure continuation of the current farming practices:

- Along roadsides — The road verges adjacent to the cropping paddocks were devoid of trees. John felt that some of the local trees and shrubs could be established along these verges by direct seeding with the approval of the district council. The technique previously used along the creekline was thought to be suitable. Seed was obtained from remnant vegetation on the property.
- In the hills — Many of the trees in the hills are old and starting to show signs of senescence. John was keen to replant individual seedlings amongst the older trees to provide shelter in the future. Seed collected from the remnant trees was used to grow 70 seedlings for the first stage of the project.

- On non-productive areas — Small areas of hard red non-productive clay were selected for planting. Following a reconnaissance of tree plantations in the vicinity of the property the following species were chosen: Flinders Range wattle, Broughton willow, grey buloke, Kangaroo Island narrow leaved mallee, salmon gum, salt river gum and red ironbark. Although none of these species are local to the area it was argued that they were best suited to the red clays which, prior to settlement, had been treeless. At the time it was TFL's policy to supply only the local native species, but due to the experimental nature of this project an exception was made. Seedlings were planted on a grid with 5 m x 5 m spacings.
- Along drainage lines — A mixture of red gums, South Australian blue gums and red ironbarks were selected for the drainage lines. Trees were widely scattered (50 m apart) along the lines.
- Around watering points — Trees were placed near troughs in the cropping paddocks to provide shade for stock.

A master plan was prepared showing areas to be planted in relation to the farm layout. Detailed designs were prepared for the red clay areas, identifying the placement of individual trees and shrubs. A three-year staged planting program was then identified, commencing in 1988. The first year entailed:

- direct seeding along 1 km of roadside on the western boundary of the property;
- planting 70 seedlings in individual guards in the hills;
- planting approximately 600 seedlings in two small areas of red clays to evaluate the suitability of the species selected.

This was an ambitious first attempt at such a program but it was felt that the assistance available would make it achievable. TFL supplied the seedlings free of charge and helped collect the seed for direct seeding. A grant from Greening Australia (SA) provided the funds to buy the materials for the tree guards. John prepared the red clay areas in advance by spraying herbicide and ripping the planting lines. In early July a team of TFL planters planted the areas in one day.

Despite this assistance, John maintains that in hindsight it was too much to take on too soon. After the planting dry weather prevailed and John was forced to water the seedlings in early summer. In an area with 500 mm average annual rainfall this should normally not be necessary. Also, further weed control around the seedlings proved to be necessary and very time consuming.

Despite the difficulties encountered the first stage has been successful. All individually planted seedlings have survived, and on the red clays there has been an 80 per cent survival rate. By January 1990, 18 months after planting, some of the seedlings on the clay were 1.3 m high. Despite the initial success the second and third stages of the planting program have been deferred until the performance of the seedlings on the red clays has been further monitored.

Establishment of seedlings along the riplines was patchy, probably due to the unreliable rainfall following seeding that year. Despite the apparent poor results along the roadside there are still many seedlings present and with time they may provide sufficient cover. Weed competition along the verges is such that it will be necessary in future to commence site preparation well in advance of seeding in order to obtain good weed control. In view of the apparent poor results along the roadside John is not prepared to use this technique again until an assured technique has been developed specifically for the roadsides.

Like many landholders who have successfully carried out tree planting projects, both large and small, John maintains that establishing trees is 'harder in practice than in theory' and argues that 'it is far better to retain the remnant native vegetation on the farm than it is to try to put it back'. For those who are not fortunate enough to have remnant vegetation on their properties they have no choice but to establish trees either by planting, natural regeneration or direct seeding. Organisations such as Trees for Life and Greening Australia provide valuable and much needed support to landholders re-establishing trees on their land for either conservation or land management purposes.

PHOTOGRAPH 17

John Smyth, 'Salter Springs', inspecting three-year-old South Australian blue gums established by direct seeding into strips treated with herbicide.

An Old Motto at Work in Land Management: Adam Clark's Story

By Peter Voller

'Live as though you will die tomorrow, but farm as though you will live forever'. This is the Clark family motto, says Adam Clark of 'Bimbadeen' near Taroom in Central Queensland.

Since 1938, Adam's family has been developing and managing their 6400 ha property for beef production, cropping, and as a stable environment for future generations. From the outset, when Adam's father first bought the blocks, the family has faced the task of opening up the eucalypt forest country, river flats and dense brigalow for farming. Fortunately Adam's family was blessed with a conservative 'hasten slowly' attitude towards land development.

As Adam says: 'there is no fast road to Utopia. Long-term benefits often have short-term costs... If people are only interested in short-term gains and present ease of running a property, then some future generation will pay for their short-sightedness.'

Across the whole of their property is a network of five-chain-wide shadelines of original vegetation. They have been laid out to help with providing shade for stock, to keep the country warmer, to act as fire breaks and to help in stock mustering. Later Adam realised the great benefit these retained areas could have for wildlife preservation.

'Bimbadeen' has become a haven for wildlife of many types. Adam, who is a founding member of the local branch of the Wildlife Preservation Society, reckons that he has not lost a single species of bird, plant or animal from his property despite the clearing that has been so necessary for his own living.

Presently about 15 per cent of the property is still uncleared. In places where there has been over-clearing in the past, regrowth is being allowed to get away and grow into new shadelines and wildlife corridors.

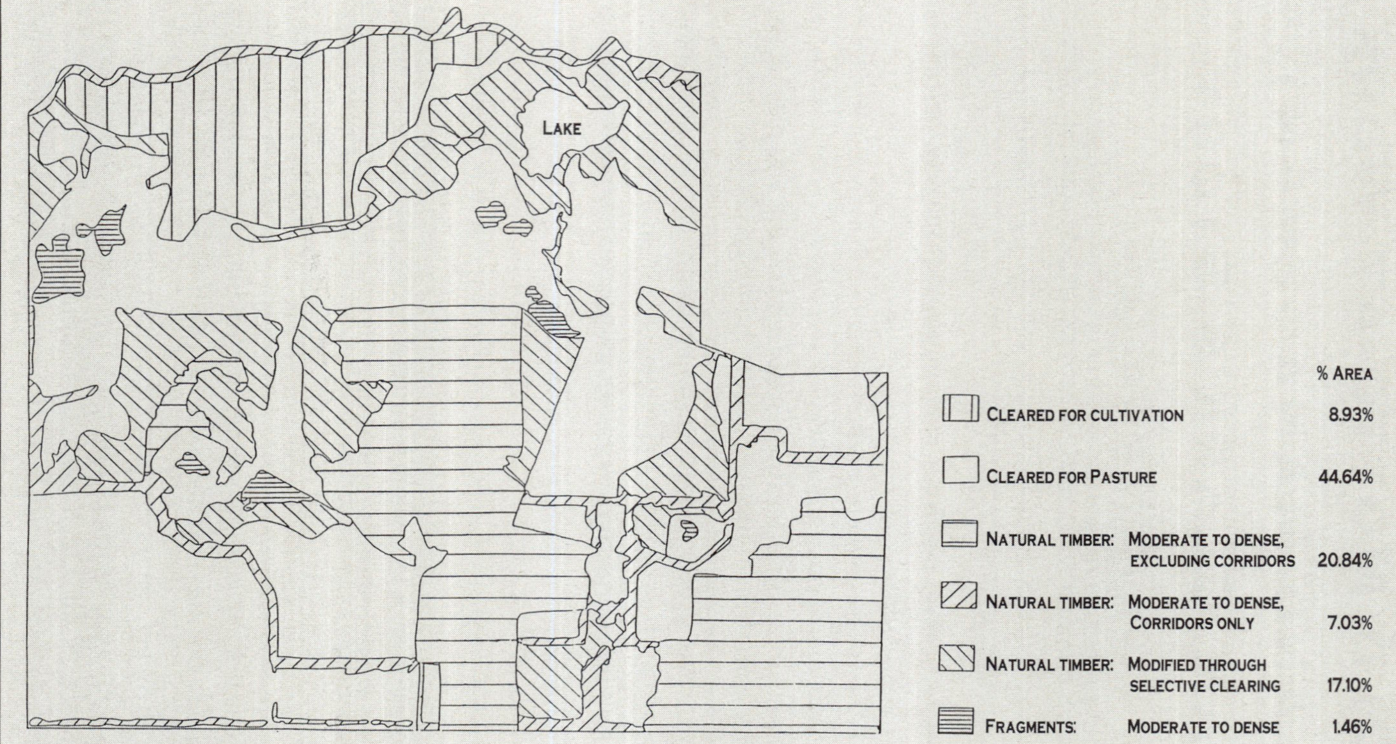

FIGURE 27

Clark family holding 'Bimbadeen' via Taroom, showing an outline of the breakup of farming and grazing units. From Lawrie (1989), by courtesy of the Queensland Department of the Environment and Heritage.

Adam and his sons, who presently work the place with him, try to look at their property as a whole living organism. All the original vegetation types have been preserved and some parts of the place are to be set aside purely for wildlife habitat, such as a patch of softwood scrub under a mountain range where ooline trees grow. Ooline (*Cadellia pentastylis*) was once reasonably common in the area, but over-clearing in some places has made it rare.

Adam sees that the only way to keep land productive, and to be able to look at other options such as wildlife preservation, is to plan ahead on a whole farm basis. In that way all aspects of human involvement can be viewed from an economic and ecological standpoint.

The Clark family draws a great deal of satisfaction from sharing their property with wildlife although they admit that, at times, wallabies and cockatoos can be pests in their crops.

'Bimbadeen' is a popular resting spot for several migratory birds including magpie geese, pelicans and the occasional jabiru who call by and stay on the 400 ha seasonal wetland area. Ducks and many other water birds are more regular inhabitants of the area. 'Habitat is the key', says Adam. 'It's not much use having wildlife if there is nowhere for it to live'. In this regard, as a land manager and long-time nature lover, Adam Clark has shown his ability to effectively run a commercial property, while at the same time work to protect the stability of his land.

PHOTOGRAPH 18

Adam Clark, whose family has been developing 'Bimbadeen' near Taroom for beef production, cropping and as a stable environment since 1938.

GLOSSARY

alimentary canal	gut, a tube running from mouth to anus
annual	plant with a one-year life cycle
araneologist	one who studies spiders
asymptotic	approaching a high point less and less rapidly
antenna (pl, antennae)	jointed, paired sensory structure attached to the head of insects and some other arthropods; feelers
arthropod	invertebrate organism with an exoskeleton, jointed appendages and segmented body e.g. insects, crustacea, millipedes, centipedes, spiders and scorpions
biological control	control of plant or animal populations by other organisms — predators, parasites and pathogens
biotic	living
biotic community	populations of plants and animals found together
carnivore	organism which eats animal flesh
chlorinated hydrocarbons	insecticides containing hydrocarbon molecules with bonded chlorine atoms e.g. DDT, endrin, BHC, lindane, chlordane, heptachlor, aldrin, dieldrin; destroy nerve tissue (nerve poisons); and affect a wide range of animals. Some are highly toxic to mammals while others have low toxicity; DDT is soluble in animal fats and persists for long periods, as do most other chlorinated hydrocarbons
compound eyes	large paired eyes of insects, comprising many small facets

crustacea	members of the Class Crustacea e.g. crabs, prawns, shrimps, barnacles, slaters (woodlice)
cuticle	thin outer covering of plants or animals; in insects and other arthropods forms a strong, protective exoskeleton
DDT	dichlordiphenyltrichloroethane; nerve poison used as an insecticide — see chlorinated hydrocarbons
decomposers	organisms involved in the physical and chemical breakdown of organic plant and animal remains
detritivore	organism which feeds on detritus (organic debris)
dragon	lizard belonging to the Family Agamidae
ecology	study of organisms in relation to their environment
economic threshold	population density of a pest above which economic loss justifies the cost of control measures
ecosystem	biotic community (plants and animals) together with its abiotic (non-living) environment e.g. a log, lake, desert, pasture, planet earth
elytra	hardened fore-wings of beetles
energy flow	the passage of energy from organism to organism, as a result of feeding processes, with some energy being lost as heat at each transition
entomophagous	feeding on insects
environment	all the biotic (living) and abiotic (non-living) factors directly affecting an organism's chance to survive and reproduce
environmental stress	stress induced by extreme environmental conditions e.g. heat, cold, dryness — possibly resulting in death, depending upon the intensity and duration of exposure

epicormic growth	growth of shoots from special buds in the bark of *Eucalyptus* trees; often occurs after stress e.g. fire
epidermis	in plants, a single layer of cells at the surface of leaves and young stems
epizootic	disease temporarily causing high mortality in an animal population
exoskeleton	external covering of some animals, either bony or leathery
extrapolate	extend trend beyond known or observed values or conditions
fecundity	potential number of offspring produced by a female in her lifetime
feral	animals in the wild state, but derived from introduced, domestic stocks e.g. feral cats, pigs, goats and camels in Australia
fungivore	organism which eats fungi
fungus (pl, fungi)	simple organisms lacking chlorophyll; many important as decomposers or parasites e.g. mushrooms and mildew
gall insects	those which induce an abnormal growth of plant tissue in which the insect shelters
germination	initial growth of a seed
gizzard	part of the gut in some animals used to crush food before digestion
habitat	place with the specific set of conditions required for a species or community to survive and reproduce; a place to live e.g. pond for dragonfly nymphs or eucalypt woodland for kookaburras
herbivore	organism which eats plants
herpetology	study of amphibians and reptiles
hibernation	winter dormancy characterised by low level of activity

honeydew	sweet secretion of some sap-feeding insects e.g. aphids
hymenopteran	insect of the Order Hymenoptera — ants, bees, wasps and sawflies
improved pasture	one sown to introduced grasses and clover, fertilised and perhaps partly or wholly cleared of native trees
indigenous	belonging to a natural ecosystem; not introduced by man; native
insect	member of Class Insecta; having one pair of antennae, three pairs of legs and body divided into head, thorax and abdomen e.g. beetles, flies, wasps, bugs
insecticide	poisonous chemical compound used to kill insect pests
insectivore	organism which eats insects — hence, insectivorous
integrated pest control	rational control of pests by combining, where necessary, biological, cultural, chemical and natural controls
interaction	strictly, the difference between the effect of a factor (e.g. nitrogen fertilisation) in the presence and absence of another factor (e.g. superphosphate fertilisation) or at different levels of each factor; but also used, less rigorously, to mean the way two or more organisms, species or processes affect each other
invertebrates	animals without backbones e.g. worms, insects
larva (pl, larvae)	immature stage of certain insects, bearing little resemblance to the adult
lerp	a protective scale constructed by some insects on a leaf or twig or insects using these e.g. psyllids

life cycle	the cyclic stages of reproduction characteristic of a species
maggot	fly larva
mandibles	in insects and some other arthropods, opposed jaws usually used to crush food, but vary in form depending on food preference
metamorphosis	rapid transformation during life cycle of some organisms e.g. pupa to butterfly
moult	in arthropods, replacement of old exoskeleton by a new one which has formed underneath
myxomatosis	virus disease of rabbits introduced to Australia from South America in 1950 as a control measure
native pasture	grassland consisting of indigenous grasses and other herbs
natural controls	naturally occurring biotic and abiotic factors which influence the abundance of animal populations e.g. predators, parasites, disease and weather
niche	way of life, role and activities of an organism in the ecosystem
nitrogen fixation	alteration of gaseous nitrogen to nitrogenous compounds by bacteria in root nodules or free-living in soil, thereby making nitrogen available to plants
nutrient cycling	cycling of nutrients through an ecosystem by physical, chemical and biological processes e.g. soil → plant → animal → soil
nymph	immature stage of certain insects, bearing some resemblance to the adult

organophosphates	synthetic compounds, originally developed as nerve gases during World War Two, and later used as insecticides e.g. parathion, dimefox, sulfotepp, TEPP, malathion, diazinon, rogor, ronnel. Some are highly toxic to mammals, producing severe dislocations in the nervous system, but several have the advantage that they do not leave long-term residues; many have systemic properties — see systemic insecticides
oxygen debt	in reptiles, depletion of oxygen in the blood due to sudden burst of activity
panting threshold	the temperature at which a reptile begins to pant in order to cool itself
parasite	organism which lives at the expense of another (the host) without necessarily causing its death e.g. flea feeding on dog, mistletoe on tree
pathogen	a parasite causing disease
pedipalps	a pair of leg-like sensory appendages between the jaws and front pair of legs in spiders
pest	an organism deemed undesirable, because it causes economic loss or ill health, or is a nuisance
pesticide	poisonous chemical compound used to attempt control of a pest
pesticide resistance	resistance to pesticides in a pest species due to selection for resistant individuals following previous applications
phytophagous	feeding on plants
poikilothermic	with body temperature approximately that of the surroundings, dependent upon external sources of heat; 'cold-blooded' (misleading) e.g. invertebrates, fish, amphibians, reptiles
pollen	powder produced by flowers, each grain containing a male gamete

pollinator organism which transfers pollen from flower to flower e.g. bees, moths

population reproductive group of organisms of the same species

population density number of individuals in a given area, volume or other unit of habitat e.g. number per square metre, number per leaf

population dynamics the changes in structure and characteristics (e.g. birth and death rates, age distribution, sex ratio and density) of a population

population regulation control of population size by action of abiotic and biotic environmental components; often defined, more specifically, as the operation of density-dependent factors (those factors which vary in intensity depending upon the population density)

predator organism which preys upon others — hence predacious

progeny offspring

prothorax anterior part of the thorax

protozoan microscopic, unicellular organism of the Phylum Protozoa

provenance place of origin

pupa inactive stage, between larva and adult, in the life cycle of certain insects

queen fertile female in colony of social insects; main function is to lay eggs

raptorial adapted for grasping prey; predatory

regeneration new growth of saplings, bushes and other plants from seed and sucker growth following disturbance

ringbarking	practice of cutting a ring of bark around a tree trunk to kill the tree, called girdling in early descriptions. Sucker growth sometimes occurs from the base of the trunk
sap	watery solution of sugars and other substances produced by plants, present in roots and shoots
sclerophyllous plants	those with hard, leathery leaves e.g. *Eucalyptus* — hence sclerophyll woodland
secondary woodland	that resulting from regeneration after disturbance of the pristine woodland community
soil micro-organisms	microscopic organisms (e.g. bacteria, fungi) which utilise organic matter (e.g. litter, dung, animal remains) and break it down to simple mineral nutrients able to be absorbed by plant roots
soil porosity	pore space between soil particles
species	group of similar organisms able to produce fertile young, if mated, and deemed to be a distinct kind by taxonomists
species diversity	the number of species within a particular place at a particular time
sting	modified ovipositor of some hymenopteran insects, able to secrete venom and used in defence and food capture; painful effects of same
sucker growth	shoots growing from roots or stumps of trees, hence 'sucker-bashing'
systemic insecticide	insecticide absorbed by plants and transported through the plant (e.g. from leaves to stems) and therefore often used against plant-sucking pests
terrestrial	living on land

territory	area of habitat defended by a pair or group of animals against other members of the same species — hence territorial
thermo-regulation	maintenance of body temperature
thorax	middle section of the insect body
topography	landscape
understorey	vegetation of a height between mature trees and ground cover e.g. saplings, small trees and shrubs
venation	system of veins in leaves or insect wings; their arrangement
virus	submicroscopic organism infecting plants or animals e.g. myxomatosis
volunteer grasses	grasses, other than those sown, appearing in pasture from seed blown in or brought in by animals, e.g. barley grass and couch grass; not necessarily weeds, because grazed by stock
weed	plant growing where it is not wanted
weedicide	chemical applied to attempt weed control
wilting point	soil moisture content at which plants wilt permanently unless soil is moistened, because the roots are unable to absorb water; soil air is no longer saturated below this point. Not to be confused with temporary wilting of plants whenever water absorption is slower than loss of water by transpiration from leaves
worker	caste of the social insects (e.g. ants, wasps, bees) which collects food, maintains the colony and tends the queen

Classification of Plants and Animals

Conspicuous and well-known plants and animals are often labelled with common names which we all know e.g. 'red gums' or 'black snakes', but all known organisms have also been formally described by taxonomists and given a binomial Latin name consisting of the genus and species. Hence, the Australian magpie has the scientific name *Gymnorhina tibicen*. It belongs to the genus *Gymnorhina* (meaning bare nose) and its specific name is *Gymnorhina tibicen* (*tibicen* meaning a flute player).

A genus consists of one or more similar species and a family of several similar genera and so on. Thus organisms are classified by means of a hierarchy of groupings according to their similarity and, perhaps, their closeness in an evolutionary sense. The groupings or taxa are ranked as follows:

Taxon	**Example:** black field cricket
Kingdom	Animalia
Phylum	Arthropoda
Class	Insecta
Order	Orthoptera
Family	Gryllidae
Genus	*Teleogryllus*
Species	*Teleogryllus commodus*

Intermediate taxa e.g. suborders or superfamilies may also be utilised

- In this book we often refer to family groups e.g. syrphids or scarabs. The former belong to the Family Syrphidae (hover flies) and the latter to the Family Scarabaeidae (scarab beetles). Both these families belong to the Phylum Arthropoda, Class Insecta but the scarabs are members of the Order Coleoptera (beetles) while the syrphids belong to the Order Diptera (flies).
- Common or vernacular names are often more colourful and immediately descriptive than the Latin equivalents, but in different countries or regions the same common name may be applied to different species or groups. Conversely, the same species may have different common names in different regions, these sometimes being remarkably parochial. In scientific circles, the use of Latin names usually avoids such confusion. However, there may be confusion with scientific names because a species may be described and named more than once, or a species may be moved from one genus to another. Strictly the scientific name is followed by the name of the person who described and named that species, usually abbreviated. Perennial ryegrass, for example, is *Lolium perenne* L., the last letter standing for Linnaeus, who first named this grass and described it.

Common and Scientific Names: Plants

Apple, rough-barked	*Angophora floribunda*
Apple box	*Eucalyptus bridgesiana*
Australian indigo	*Indigofera australis*
Bamboo grass, stout	*Stipa ramosissima*
Barley grass	*Hordeum leporinum*
Biddy bush, sifting bush	*Cassinia quinquefaria*
Blackberry	*Rubus procerus* (= *R. vulgaris*)
Black sally	*Eucalyptus stellulata*
Blackthorn, native box	*Bursaria spinosa*
Blue gum	*Eucalyptus leucoxylon*
Blue mallet	*Eucalyptus gardneri*
Box	
apple	*Eucalyptus bridgesiana*
peppermint	*Eucalyptus odorata*
yellow	*Eucalyptus melliodora*
Broughton willow, coobah	*Acacia salicina*
Cedar, red	*Toona australis*
Cocksfoot	*Dactylis glomerata*
Coral gum	*Eucalyptus torquata*
Daisies, daisy bushes	Asteraceae
biddy bush	*Cassinia quinquefaria*
sticky daisy bush	*Olearia elliptica*
tree everlasting	*Helichrysum dendroideum*
white paper daisies	*Helichrysum elatum*
yellow buttons	*Helichrysum apiculatum*
Dogwood	*Jacksonia scoparia*

Eggs and bacon, spreading bush pea	*Pultenaea microphylla*
Epacrids	Epacridaceae
blunt-leaf heath	*Epacris obtusifolia*
honey heath	*Melichrus urceolatus*
peach heath	*Lissanthe strigosa*
Fern-leaf wattle	*Acacia filicifolia*
Fescue	*Festuca arundinacea*
Flatweed	*Hypochaeris radicata*
Flinders Range wattle	*Acacia iteaphylla*
Golden wattle	*Acacia pycnantha*
Gorse bitter-pea, native gorse	*Daviesia ulicifolia*
Grey buloke, swamp she-oak	*Casuarina glauca*
Gum	
Blakely's red	*Eucalyptus blakelyi*
blue	*Eucalyptus leucoxylon*
coral	*Eucalyptus torquata*
manna, ribbon, white	*Eucalyptus viminalis*
mountain white	*Eucalyptus dalrympleana*
river red, Murray red	*Eucalyptus camaldulensis*
salmon, yellow	*Eucalyptus salmonophloia*
salt river	*Eucalyptus sargentii*
snow	*Eucalyptus pauciflora*
Strickland	*Eucalyptus stricklandii*
white	*Eucalyptus viminalis*
Hickory	*Acacia implexa*
Honey heath	*Melichrus urciolatus*

Hopbush	*Dodonaea* spp.
Ironbark	
narrow-leaved red	*Eucalyptus crebra*
red, mugga	*Eucalyptus sideroxylon*
Kangaroo grass	*Themeda triandra*
Kangaroo Island	
narrow leaved mallee	*Eucalyptus cneorifolia*
Legumes	
Mimosaceae — mostly acacias	*Acacia* spp.
Caesalpinaceae —	
rare on tablelands	*Cassia* spp.
Fabaceae — many genera	
Australian indigo	*Indigofera australis*
bitter-pea	*Daviesia* spp.
dogwood	*Jacksonia scoparia*
eggs and bacon	*Pultenaea* spp.
happy wanderer	*Hardenbergia violacea*
Lovegrass	*Eragrostis* spp.
Manna gum	*Eucalyptus viminalis*
Mallet	
blue	*Eucalyptus gardneri*
swamp	*Eucalyptus spatulata*
Mountain white gum	*Eucalyptus dalrympleana*
Mount Leseuer mallee	*Eucalyptus suberea*
Narrow-leaved red ironbark	*Eucalyptus crebra*
Native box	*Bursaria spinosa*
New England peppermint	*Eucalyptus nova-anglica*
Nodding thistle	*Carduus nutans*
Ooline	*Cadellia pentastylis*

Peach heath	*Lissanthe strigosa*
Perennial ryegrass	*Lolium perenne*
Phalaris	*Phalaris aquatica*
Red gum	
Blakely's	*Eucalyptus blakelyi*
river	*Eucalyptus camaldulensis*
Red ironbark	*Eucalyptus sideroxylon*
Red-leg grass, red grass	*Bothriochloa macra*
Rough-barked apple	*Angophora floribunda*
Ryegrass	
annual, Italian	*Lolium multiflorum*
perennial	*Lolium perenne*
Saffron thistle	*Carthamus lanatus*
Salt river mallet	*Eucalyptus sargentii*
Salmon gum	*Eucalyptus salmonophloia*
Shepherd's purse	*Capsella bursa-pastoris*
Silver grass	*Vulpia* spp.
Snow grass, fine-leaf tussock	*Poa sieberiana*
Snow gum	*Eucalyptus pauciflora*
Spear thistle	*Cirsium vulgare*
Sticky daisy bush	*Olearia elliptica*
Stout bamboo grass	*Stipa ramosissima*
Strickland gum	*Eucalyptus stricklandii*
Stringybark	
red	*Eucalyptus macrorhyncha*
silvertop	*Eucalyptus laevopinea*
New England	*Eucalyptus caliginosa*
Stylo	*Stylosanthes* spp.
Swamp mallet	*Eucalyptus spathulata*

Sweet briar	*Rosa rubiginosa*
Tea trees	*Leptospermum* spp.
Thistles	
nodding	*Carduus nutans*
saffron	*Carthamus lanatus*
spear	*Cirsium vulgare*
Traveller's joy	*Clematis aristata*
Urn heath	*Melichrus urceolatus*
Wallaby grass	*Danthonia* spp.
White clover	*Trifolium repens*
White gum	*Eucalyptus viminalis*
White paper daisies	*Helichrysum elatum*
Yellow buttons	*Helichrysum apiculatum*

Common and Scientific Names: Animals

Antlion, antlion fly	Myrmeleontidae
Aphid, rose aphid	*Macrosiphum rosae*
Armoured scale	Diaspididae
Assassin bug	Reduviidae
Australian magpie	*Gymnorhina tibicen*
Bearded dragon	*Amphibolurus barbatus*
Bee fly	Bombyliidae
Bittern	*Botaurus poiciloptilus*
Black field cricket	*Teleogryllus commodus*
Black soil scarab	*Othnonius batesi*
Black swan	*Cygnus atratus*
Blue tongued lizard	*Tiliqua scincoides*
Borer (longicorn beetle)	Cerambycidae
Broad-nosed bat	*Nycticeius* spp.
Brown hare	*Lepus capensis*
Butcher bird	*Cracticus torquatus*
Cabbage white butterfly	*Pieris rapae*
Case moth	Psychidae
Cherry slug	Tenthredinidae
Christmas beetle	*Anoplognathus* spp.
Click beetle	Elateridae
Corbie (grassgrub)	*Oncopera* spp.
Crimson rosella	*Platycercus elegans*
Crow	*Corvus* spp.
Cuckoo shrike	*Coracina novaehollandiae*

Cup moth	Limacodidae e.g. *Doratifera* spp.
Curl grub (scarab larva)	Scarabaeidae
Currawong	*Strepera* spp.
Cutworm	Noctuidae e.g. *Agrotis, Spodoptera*
Dabchick, little grebe	*Tachybaptus novaehollandiae*
Daddy-long-legs	*Pholcus phalangiodes* (Pholcidae)
Damselfly	Zygoptera
Dingo	*Canis familiaris dingo*
Double-barred finch	*Poephila bichenovii*
Dragonfly	Anisoptera
Dusky pasture scarab	*Sericesthis nigrolineata*
Eastern brown snake	*Pseudonaja textilis*
Eastern grey kangaroo	*Macropus giganteus*
Eastern rosella	*Platycercus eximius*
Eastern tiger snake	*Notechis scutatus*
Echidna	*Tachyglossus aculeatus*
Emperor gum moth	*Antheraea eucalypti*
Flower spider	Thomisidae
Flower wasp	Tiphiidae (thynnids) e.g. *Hemithynnus* spp.
Fox	*Vulpes vulpes*
Funnel-web spider	*Atrax* spp.
Galah	*Cacatua roseicapilla*
Gall aphid	Phylloxeridae
Gould's wattle bat	*Chalinolobus gouldii*
Grey fantail	*Rhipidura fuliginosa*
Ground beetle	Carabidae

Gum tree scale	*Eriococcus coriaceus*
Hairy flower wasp	Scoliidae e.g. *Scolia* spp.
Honeyeater	*Lichenostomus* spp.
Hover fly	Syrphidae
Huntsman spider	Sparassidae
Jacky lizard	*Amphibolurus muricatus*
Jumping spider	Salticidae
Kookaburra	*Dacelo novaeguineae*
Lacewing	Neuroptera
Ladybird	Coccinellidae
Lapwing, spur-winged plover	*Vanellus miles*
Leaf beetle	Chrysomelidae e.g. *Paropsis*
Leaf miner	Moth larvae e.g. Incurvariidae
Leaf roller spider	*Phonognatha* spp.
Leaf skeletoniser	Nolidae e.g. *Uraba* spp.
Leaf tier	Tineoidea
Legless lizard	Pygopodidae
Lerp	Psyllidae e.g. *Cardiaspina* spp.
Little grebe	*Tachybaptus novaehollandiae*
Long-eared bat	*Nyctophilus geoffroyi*
Lynx spider	Oxyopidae
Magpie-lark	*Grallina cyanoleuca*
Mantispid	Mantispidae
Mastiff-bat	Molossidae
Millipede	Diplopoda
Mistletoe-bird	*Dicaeum hirundinaceum*
Naked scale, soft scale	Coccidae
Nematode worm	Nematoda e.g. Mermithidae
New Zealand grass grub	*Costelytra zealandica*

Noisy friarbird	*Philemon corniculatus*
Noisy miner	*Manorina melanocephala*
Nursery-web spider	Pisauridae
Orb-weaver spider	Araneidae
Paper wasp	Vespidae
Pardalote	*Pardalotus* spp.
Pasture white grub	*Rhopaea* spp.
Pied cormorant	*Phalacrocorax varius*
Plague locust	*Chortoicetes terminifera*
Plover, lapwing	*Vanellus miles*
Pruinose scarab	*Sericesthis geminata*
Platform spider	Agelenidae
Praying mantis, mantid	Mantodea
Psyllid	*Cardiaspina* spp.
Raven	*Corvus* spp.
Red-back spider	*Latrodectus mactans*
Red bellied black snake	*Pseudechis porphyriacus*
Red wattlebird	*Anthochaera carunculata*
Robber fly	Asilidae
Rove beetle	Staphylinidae
Sawfly	*Perga* spp.
Scale insect	*Eriococcus coriaceus*
Scarab	Scarabaeidae
Scelionid wasp	Scelionidae e.g. *Scelio* spp.
Scissors grinder, restless flycatcher	*Myiagra inquieta*
Scorpion	Scorpionida
Scorpion-tailed spider	*Arachnura* spp.
Silvereye	*Zosterops lateralis*

Skink	*Lampropholis* spp.
Soldier bird	*Manorina melanocephala*
Spinebill	*Acanthorhyncus tenuirostris*
Spiny anteater	*Tachyglossus aculeatus*
Springtail	Collembola
Spur-throated locust	*Austracris guttulosa*
Starling	*Sturnus vulgaris*
Stick insect	Phasmatidae
Strawnecked ibis	*Threskiornis spinicollis*
Sugar glider	*Petaurus breviceps*
Superb fairy wren	*Malurus cyaneus*
Tableland pasture scarab	*Antitrogus morbillosus*
Tachinid fly	Tachinidae
Termite	Isoptera
Thrips	Thysanoptera
Tortoise beetle	Chrysomelidae e.g. *Paropsis*
Trap-door spider	Ctenizidae
Underground grassgrub	*Oncopera* spp.
Vedalia ladybird	*Rodolia cardinalis* (Coccinellidae)
Wattlebird	*Anthochaera carunculata*
Weevil	Curculionidae
White ibis	*Threskiornis molucca*
Willy wagtail	*Rhipidura leucophrys*
Wingless grasshopper	*Phaulacridium vittatum*
Wire-worm	Elateridae
Wolf spider	Lycosidae
Yellow-rumped thornbill	*Acanthiza chrysorrhoa*
Yellow-tailed black cockatoo	*Calyptorhynchus funereus*
Yellow-bellied grasshopper	*Praxibulus laminatus*

References and Further Reading

Abbott, W. (1880), 'Ringbarking and its effects', *Journal Proceedings Royal Society of New South Wales* 14: 97–102.

Abercrombie, M., Hickman, C. and Johnson, M. (1977), *The Penguin dictionary of biology*, Allen Lane, London.

Baker, G. (1981), 'Wingless grasshopper', NSW Agriculture Agfact AE.I.

Baker, G. (1983), 'Parasites of locusts and grasshoppers', NSW Agriculture Agfact AE.2.

Baker, G. (in press), 'Current outbreak and pest status of wingless grasshopper, *Phaulacridium vittatum*, in south-eastern Australia', in *Pests of pastures: weed, invertebrate and disease pests of Australian sheep pastures*, ed. E. Delfosse, CSIRO Publications, Melbourne.

Barker, R. and Vestjens, W. (1979), *The food of Australian birds*, CSIRO Publications, Melbourne.

Beckmann, R. (1989), 'Rural dieback: restoring a balance', *Ecos* 62: 9–15.

Beckmann, R. and Davidson, S. (1990), 'Reversing rural tree decline', *Rural Research* 146: 19–26.

Begg, J. (1959), 'Annual pattern of soil moisture stress under sown and native pasture', *Australian Journal Agricultural Research* 10: 518–29.

Bird, P., Bicknell, D., Bulman, P., Burke, S., Leys, J., Parker, J., Van der Sommen, F. and Voller, P. (in press), 'The role of shelter in Australia for protecting soils, plants and livestock', *Agroforestry Systems Journal*.

Bowen, S. (1981), Factors contributing to the maintenance of an ecotone between a woodland and a grassland on the northern tablelands of NSW, PhD thesis, University of New England.

Boyd, K. and Phipps, R. (1978), Occurrence of magpies in relation to distance from trees, ecology project, University of New England.

Boyd, R. (1976), *Mortality of the native trees in the woodland ecosystems of the New England tablelands of NSW and related topics*, Extension Services Division, University of New England.

Breckwoldt, R. (1983), *Wildlife in the home paddock*, Angus and Robertson.

Breckwoldt, R. (1986), *The last stand: managing Australia's remnant forests and woodlands*, AGPS, Canberra.

Breckwoldt, R., Brown, I., Collins, M., Gorrie, I., Hussey, P., Johnston, P., Lewis, S., Loney, B., Middleton, W., Parker, N., Turner, H. (1990), *Living corridors: conservation and management of roadside vegetation*, Greening Australia, Canberra.

Burn, A., Coaker, T. and Jepson, P. (1987), *Integrated pest management*, Academic Press, London.

Caley, N. (1984), *What bird is that?*, rev. T. Lindsay, Angus and Robertson, Sydney.

Campbell, A. (1991), *Planning for sustainable farming: the Potter farmland plan story*, Lothian, Melbourne.

Campbell, A. (1991), *Insect pests of farm trees*, no.8 in Farm Trees Series, NSW Agriculture, Forestry and Soil Conservation.

Carne, P. (ed.) (1987), *Scientific and common names of insects and allied forms occurring in Australia*, CSIRO Publications, Melbourne.

Carne, P., McInnes, R. and Green, J. (1980), 'Seasonal fluctuations in the abundance of two leaf-eating insects', in *Eucalypt dieback in forests and woodlands*, eds K. Old, G. Kile & C. Ohmart, CSIRO Publications, Melbourne.

Carrick, R. (1959), 'The food and feeding habits of the straw-necked ibis, *Threskiornis spinicollis* (Jameson), and the white ibis, *T. molucca* (Cuvier), in Australia', *CSIRO Wildlife Research* 4: 69–92.

Carrick, R. (1962), 'Breeding, movements and conservation of ibises (Threskiornithidae) in Australia', *CSIRO Wildlife Research* 7: 71–88.

Carrick, R. (1963), 'Ecological significance of territory in the Australian magpie, *Gymnorhina tibicen*', *Proceedings of XIII International Ornithological Congress*.

Commonwealth Scientific and Industrial Research Organization, Division of Entomology (1991), *The insects of Australia: a textbook for students and research workers*, eds I. Naumann, P. Carne, J. Lawrence, E. Nielsen, J. Spradbery, R. Taylor, M. Whitten & M. Littlejohn, Melbourne University Press, Carlton.

Conacher, J. (1989), *Pests, predators and pesticides: some alternatives to synthetic pesticides*, Organic Growers Association, Wembley, WA.

Cook, M. and Malecky, J. (1974), *Economic aspects of pasture improvement*, Wool economic research report 25, Bureau of Agricultural Economics, Canberra.

Crome, B. (1981), 'The diet of some ground-layer lizards in three woodlands of the New England tablelands of Australia', *Herpetofauna* 13: 4–11.

Cremer, K. (1990), *Trees for rural Australia*, Inkata Press, Melbourne.

Curtis, D. (1989), Eucalypt re-establishment on the northern tablelands of NSW, MSc thesis, University of New England.

Curtis, D. (1990), 'Dieback — where to now?', *Acres Australia* 3: 9–12.

Curtis, D. (1991), 'Farm trees on the northern tablelands — the next decade', *Australian Journal Soil and Water Conservation* 4: 9–15.

Davidson, R. (1982), 'An ecological view of pasture pests on the northern tablelands of NSW', *Proceedings 3rd Australasian Conference Grassland Investebrate Ecology*, ed. K. Lee, South Australian Government Printer, pp. 67–79.

Davidson, R. (1982), 'Correcting past mistakes — loss of habitat for predators and parasites of pasture pests', *Proceedings 3rd Australasian Conference Grassland Ecology*, ed. K. Lee, South Australian Government Printer, pp. 199–206.

Davidson, R., Wensler, R. and Wolfe, V. (1970), 'Damage to ryegrass plants of different size by various densities of the pruinose scarab (*Sericesthis geminata* (Coleoptera))', *Australian Journal Experimental Agriculture and Animal Husbandry* 10: 166–71.

Davidson, S. (1986), An experimental analysis of fecundity and mortality in the black field cricket, *Teleogryllus commodus* (Orthoptera: Gryllidae), MSc thesis, University of New England.

Davidson, S. (1989), 'Tree clearing in the semi-arid tropics', *Rural Research* 144: 24–8.

Department of the Arts, Heritage and the Environment and the Institute of Foresters of Australia (1985), *Think trees, grow trees*, eds I. Gorrie, A. Brown, W. Crane & R. Florence, AGPS, Canberra.

Department of the Prime Minister and Cabinet (1992), *CRC compendium*, AGPS, Canberra.

Dethier, V. (1976), *Man's plague? Insects and agriculture*, Darwin Press, Princeton, NJ.

Dixon, J. (1983), 'Gould's wattled bat', in *The complete book of Australian mammals*, ed. R. Strahan, Angus and Robertson, Sydney.

Dunkerley, G. (1989), The fuscous honeyeater: food resources and the bird community, PhD thesis, University of New England.

Emery, R. (1990), 'Wingless grasshoppers and their control', Farmnote No. 62/90, Western Australian Department of Agriculture, Agdex 622.

Farrow, R. and Baker, G. (in press), 'The reduction of wingless grasshopper infestations by sound farm management practices', in *Pests of pastures: weed, invertebrate and disease pests of Australian sheep pastures*, ed. E. Delfosse, CSIRO Publications, Melbourne.

Floyd, R. and Woodland, D. (1981), 'Localisation of soil-dwelling scarab larvae by the black-backed magpie *Gymnorhina tibicen* (Latham)', *Animal Behaviour* 29: 510–17.

Ford, H. (1986), *Ecology of birds: an Australian perspective*, Surrey Beatty & Sons, Sydney.

Frith, H.(ed.) (1976), *Birds in the Australian high country*, Reed, Sydney.

Gall, J. and Davidson, S. (1981), 'Eucalypt dieback — symptom of an ailing ecosystem', *Habitat Australia* 9: 19–21.

Gaynor, S. (1991), *The northern tablelands tree establishment program*, NSW Department of Conservation and Land Management, Armidale.

Gillard, P., Williams, J. and Monypenny, R. (1989), 'Tree-clearing in Australia's semi-arid tropics: production, economics and long-term hydrological responses', *Agricultural Science* 2: 34–9.

Goldsmith, E. (1980), 'Under control?' *Ecologist* 10: 102–9.

Goode, J. (1987), *Insects of Australia*, Angus and Robertson.

Goodyer, G. (1985) 'Scarab grubs in northern tableland pastures', NSW Agriculture Agfact P2. AE.l.

Hadlington, P. and Johnston, J. (1988), *Australian trees: a guide to their care and repair,* New South Wales University Press, Kensington.

Healy, A. and Smithers, C. (1978), *Australian insects in colour,* Reed, Sydney.

Heatwole, H. and Lowman, M. (1986), *Dieback: death of an Australian landscape,* Reed, Sydney.

Hilder, J. (1963), 'Performance of several grasses in swards', *CSIRO Field Station Record* 2: 9–24.

Hilditch, J. (1974), Effects of soil temperature and soil moisture on the abundance and distribution of pasture scarabs (Coleoptera: Scarabaeidae), PhD thesis, University of New England.

Hockings, F. (1980), *Friends and foes of Australian gardens,* Reed, Sydney.

Hofler, J. (ed.) (nd), *Focus on farm trees 2: Reversing the decline of trees,* Canberra Printing and Publishing Co.

Hoult, E. (1970), Ecological investigations of the reactions of trees to drought, MSc thesis, University of New England.

Hutchinson, K. (1971), 'Productivity and energy flow in grazing fodder conservation systems', *Herbage Abstracts* 41: 1–10.

Hutchinson, K. (in press), 'The grazing resource', *Proceedings of 6th Australian Agronomy Conference, Armidale 1992.*

Ive, J. and Cocks, D. (1989), 'Rural land degradation in Australia: the role of trees in ecological solutions', *The Australian Conservation Farmer* 1:18–22.

Johnston, P. and Don, A. (1990), *Grow your own wildlife: how to improve your local environment,* Greening Australia, Canberra.

Kain, W., Thomson, W., Plunkett, D., Atkinson, D. and Malcolm, J. (1982), 'Development and implementation of integrated management of pasture pests in intensive dairy farms of the southern North Island region of New Zealand', *Proceedings 3rd Australasian Conference Grassland Ecology,* ed. K. Lee, South Australian Government Printer, pp. 339–348.

King, K. (1991), 'The role of microarthropods in decomposition and nutrient cycling in temperate pastures', abstract in *Australian Journal Ecology* 16: 131–4.

Landsberg, J. and Wylie, F. (1988), 'Dieback of rural trees', *Australia Geojournal* 17: 231–7.

Landsberg, J., Morse, J. and Khanna, P. (1990), 'Tree dieback and insect dynamics in remnants of native woodlands on farms', *Proceedings Ecological Society Australia* 16: 149–65.

Langlands, J. and Bennett, I. (1973), 'Stocking intensity and pastoral production: changes in soil and vegetation of a sown pasture grazed by sheep at different stocking rates', *Journal Agricultural Science* 81: 193–204.

Lawrie, B. (1989), 'Land development and wildlife conservation: seeking common ground', in *The management of eucalypt country in North Queensland*, Dalrymple Landcare Committee, Charters Towers.

Lines, W. (1991), *Taming the great south land: a history of the conquest of nature in Australia*, Allen and Unwin.

Lowman, M. and Heatwole, H. (in press), 'New England dieback in Australia and subsequent landscape amelioration', *Proceedings of Pacific Science Association Congress, Honolulu, 1991*.

Main, B. (1976), *Spiders*, Collins Publishers, Sydney.

McDonald, G. (1968), 'Recent pasture development on the northern tablelands of New South Wales', *Australian Geographer* 10: 382–91.

Milner, R. (1977), 'The role of disease during an outbreak of *Oncopera alboguttata* Tindale and *O. rufobrunnea* Tindale (Lepidoptera: Hepialidae) in the Ebor/Dorrigo region of NSW', *Journal Australia Entomological Society* 16: 21–6.

Nadolny, C. (1984), Eucalypt dieback on the northern tablelands of New South Wales, MSc thesis, University of New England.

New, T. (1991), *Insects as predators*, New South Wales University Press.

Newman, P., Neville, S. and Duxbury, L. (eds) (1988), *Case studies in environmental hope*, Environmental Protection Authority for the Western Australian State Conservation Strategy.

Nivison, A. (1963), 'The study of the development of oversown pastures on "Mirani"', *Australian Journal of Science* 26: 384.

Norton, A. (1886), 'On the decadence of Australian forests', *Proceedings Royal Society Queensland* 3: 15–22.

Oates, N., Greig, P., Hill, D., Langley, P. and Reid, A. (eds) (nd), *Focus on farm trees 1: The decline of trees in the rural landscape*, Capitol Press, Melbourne.

Offor, T. and Watson, R. (eds) (1991), *Growback 91*, Growback Publications, Fitzroy.

Parry, V. (1970), *Kookaburras*, Lansdowne Press (Weldon Publishers, Willoughby).

Pimentel, D. and Perkins, J. (1980), 'Pest control: cultural and environmental aspects', *American Association Advancement Science Symposium 43*, Westview Press, Boulder, Colorado.

Plumb, T.(ed.) (1982), *Atlas of Australian Resources*, 3rd series, vol.III, *Agriculture*, Division of National Mapping, Canberra.

Pottinger, P. (1977), 'Pests of grassland and small seeds crops', in *Pastures and pasture plants*, ed. R. Langer, Reed, Wellington, NZ.

Ralph, W. (1991), 'Insect resistant eucalypts', *Rural Research* 150: 28–9.

Ralph, W. (1992), 'Integrating farm and conservation priorities', *Rural Research* 150: 13–18.

Recher, H. and Holmes, R. (1985), 'Foraging ecology and seasonal patterns of abundance in a forest avifauna' in *Birds of eucalypt forests and woodlands*, ed. J. Keast, Surrey Beatty & Sons, Chipping Norton, in association with Royal Australian Ornithological Union.

Richards, G. (1983), 'Broad-nosed bats', in *The complete book of Australian mammals*, ed. R. Strahan, Angus and Robertson, Sydney.

Ridsdill Smith, J. (1970), 'The behaviour of *Hemithynnis hyalinatus* (Hymenoptera: Tiphiidae), with notes on some other Thynninae', *Journal Australian Entomological Society* 9: 196–208.

Ridsdill Smith, J. (1977), 'Effects of root-feeding by scarabaeid larvae on growth of perennial ryegrass plants', *Journal Applied Ecology* 14: 73–80.

Riemer, N. (ed.) (1986), *Trees: why we need them countrywide*, ABC Enterprises, Sydney.

Roberts, R., Ridsdill Smith, J., Porter, M. and Sawtell, N. (1982), 'Fluctuations in abundance of pasture scarabs over an 18-year period of light-trapping', *Proceedings 3rd Australasian Conference Grassland Invertebrate Ecology*, ed. K. Lee, South Australian Government Printer.

Russell, J. and Isbell, R. (eds) (1986), *Australian soils: the human impact*, University of Queensland Press, Brisbane.

Senate Select Committee on Agricultural and Veterinary Chemicals in Australia (1990), *Report*, Senate Publications Unit, Canberra.

Sexton, M. and Stevens, A. (1983), *1982 Australian plague locust control campaign*, Agriculture Protection Board, Western Australia.

Shaw, N., McInnes, S., Reid, A. and Wheeler, W. (1983), *Birds of south-eastern Australia farmlands*, Gould League of Victoria.

Sinden, J., Jones, A. and Fleming, P. (1983), *Relationships between eucalypt dieback and farm income, stocking rate and land value in southern New England, New South Wales*, University of New England Press, Armidale.

Stuckey, I. (1941), 'Seasonal growth of grass roots', *American Journal of Botany* 28: 486–91.

Suckling, G. (1983), 'Sugar glider', in *The complete book of Australian mammals*, ed. R. Strahan, Angus and Robertson, Sydney.

van den Bosch, R. (1980), *The pesticide conspiracy*, Prism Press, Dorset, UK.

Venning, J. (1988), *Growing trees for farms, parks and roadsides*, Lothian Publishing, Melbourne.

Vestjens, W. and Carrick, R. (1974), 'Food of the black-backed magpie *Gymnorhina tibicen*', *Australian Wildlife Research* 1: 71–83.

Walden, K. (1990), 'Australian plague locust', Farmnote No. 63/90, Western Australian Department of Agriculture, Agdex 612.

Williams, J. (1963), 'Studies of environment in northern New South Wales', in *New England essays*, University of New England Press, Armidale.

Willoughby, W. and Davidson, R. (1979), 'Use, management and conservation', in *Grassland ecosystems of the world*, ed. R. Coupland, Cambridge University Press, London.

Wright, D. (1987), 'Analysis of the development of major plagues of the Australian plague locust *Chortoicetes terminifera* (Walker) using a simulation model', *Australian Journal Ecology* 12: 423–37.

Wright, P. (1964), 'Pasture improvement in New England', *Armidale and District Historical Society Journal and Proceedings* 7: 15–23.